Advanced Mathematics for Practicing Engineers

The Artech House Microwave Library

Introduction to Microwaves by Fred E. Gardiol
Microwaves Made Simple: Principles and Applications by W. Stephen Cheung and Frederic H. Levien
Microwave Tubes by A. S. Gilmour, Jr.
Electric Filters by Martin Hasler and Jacques Neirynck
Nonlinear Circuits by Martin Hasler and Jacques Neirynck
Microwave Technology by Erich Pehl
Receiving Systems Design by Stephen J. Erst
Microwave Mixers by Stephen A. Maas
Feedback Maximization by B.J. Lurie
Applications of GaAs MESFETs by Robert Soares, et al.
GaAs Processing Techniques by Ralph E. Williams
GaAs FET Principles and Technology by James V. DiLorenzo and Deen D. Khandelwal
Dielectric Resonators, Darko Kajfez and Pierre Guillon, eds.
Modern Spectrum Analyzer Theory and Applications by Morris Engelson
Design Tables for Discrete Time Normalized Lowpass Filters by Arild Lacroix and Karl-Heinz Witte
Microwave Materials and Fabrication Techniques by Thomas S. Laverghetta
Handbook of Microwave Testing by Thomas S. Laverghetta
Microwave Measurements and Techniques by Thomas S. Laverghetta
Principles of Electromagnetic Compatibility by Bernhard E. Keiser
Linear Active Circuits: Design and Analysis by William Rynone, Jr.
The Design of Impedance-Matching Networks for Radio-Frequency and Microwave Amplifiers by Pieter L.D. Abrie
Microwave Filters, Impedance Matching Networks, and Coupling Structures by G.L. Matthaei, Leo Young, and E.M.T. Jones
Analysis, Design, and Applications of Fin Lines by Bharathi Bhat and Shiban Koul
Microwave Engineer's Handbook, 2 vol., Theodore Saad, ed.
Microwave Integrated Circuits, Jeffrey Frey and Kul Bhasin, eds.
Computer-Aided Design of Microwave Circuits by K.C. Gupta, Ramesh Garg, and Rakesh Chadha
Microstrip Lines and Slotlines by K.C. Gupta, R. Garg, and I.J. Bahl
Advanced Mathematics for Practicing Engineers by Kurt Arbenz and Alfred Wohlhauser
Microstrip Antennas by I.J. Bahl and P. Bhartia
Antenna Design Using Personal Computers by David M. Pozar
Microwave Circuit Design Using Programmable Calculators by J. Lamar Allen and Max Medley, Jr.
Stripline Circuit Design by Harlan Howe, Jr.
Microwave Transmission Line Filters by J.A.G. Malherbe
Electrical Characteristics of Transmission Lines by W. Hilberg
Microwave Diode Control Devices by Robert V. Garver
Tables for Active Filter Design by Mario Biey and Amedeo Premoli
Ferrite Control Components, 2 vol., Lawrence Whicker, ed.
Microwave Remote Sensing, 3 vol., by F.T. Ulaby, R.K. Moore, and A.K. Fung

Advanced Mathematics for Practicing Engineers

Kurt Arbenz
Alfred Wohlhauser

Artech House

ARTECH HOUSE, INC.
685 Canton Street
Norwood, MA 02062

International Standard Book Number: 0-89006-189-0
Library of Congress Catalog Card Number: 86-71081

Translation of *Méthodes mathématiques pour l'ingénieur,* vols. I, II, and III, originally published in French by the Presses Polytechniques Romandes, Lausanne, Switzerland. © 1983, 1981, 1982.

86 87 10 9 8 7 6 5 4 3 2 1

Preface

The primary objective of this text is to introduce the student to the fundamental concepts of the mathematical methods used in engineering applications. The material in this book comes from a sequence of courses that the senior author has offered in the Electrical Engineering Department at the Swiss Federal Institute of Technology in Lausanne. The topics were chosen according to frequency of occurrence in typical electrical engineering applications, taking into account new ideas for presentation of the subject matter. Subsequently, the courses have also proved to be of interest to students in other departments.

As currently structured, there are four courses corresponding to the four distinct parts of this book: the first covers ***numerical methods***, which today are a necessary part of a modern engineering curriculum; the second covers ***vector analysis***, which is essential to the student in his other studies in physics; the third deals with ***Fourier series*** and ***Laplace transforms***, which are probably the most important tools for the engineer; and, finally, the fourth deals with ***complex variables***, which increasingly find applications in many engineering fields, such as fluid mechanics, *et cetera.*

To facilitate the use of this book, the four parts are completely independent of one another and the individual chapters are as short as possible, being introduced by theoretical results with demonstration and followed by typical engineering applications. A course in differential and integral calculus is the sole prerequisite.

The authors have the pleasure to acknowledge the fine cooperation of the staff of Artech House for their help in translating this text from the French, originally published as *Méthodes mathématiques pour l'ingénieur; analyse numérique, compléments d'analyse, et variables complexes,* Presses Polytechniques Romandes.

K. ARBENZ

A. WOHLHAUSER

Lausanne

April 1986

Contents

Part I
Numerical Methods

Chapter 1

The Least-Squares Method

1.1 INTRODUCTION

Let us consider a ball which rolls at a constant speed v. The relationship between the time t and the position y of the ball is given by the linear equation:

$$y = a + vt \tag{1.1}$$

In measuring y for different values of t, we obtain, for example, table 1.1, which gives us the graph of figure 1.2

Table 1.1

t	0	3	4	7	8	10
y	1	4	5	6	7	10

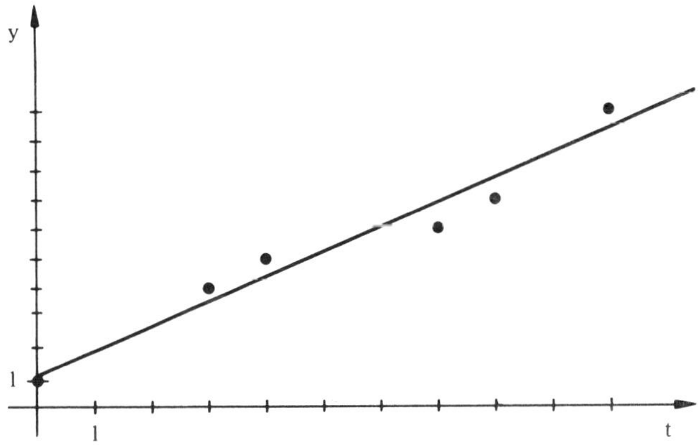

Fig. 1.2

We find that the points are only approximately located on a line. There are two explanations for this phenomenon:

- there are measurement errors,
- the speed v is probably not constant.

We do not wish to deal with the second source of error. Therefore, we assume that if y and t could be measured exactly, their relationship would be that given by equation (1.1)

Because measurements are never completely precise, a and v cannot be determined exactly, but we may try to determine the "best" values of a and v by fitting a straight line through the measured points. The slope v and the vertical offset a of the best linear fit will then be the best estimates, which are written $\hat{a}$ and $\hat{v}$.

Given that $y = y_i$ is the value of y measured at time $t = t_i$, we have the following relationship between y_i and t_i:

$$y_i = a + vt_i + r_i, \quad i = 1, 2, \ldots, m \tag{1.2}$$

where r_i is the measurement error.

Equations (1.2) are called ***measurement equations***.

The ***least-squares method*** consists now of determining a and v such that the sum of the squares of the errors is a minimum. Therefore we wish to minimize

$$S = \sum_{i=1}^{m} r_i^2 = \sum_{i=1}^{m} (y_i - a - vt_i)^2 \tag{1.3}$$

where m is the total number of observations.

From a geometric point of view, the errors r_i are the vertical distances between the measurement points and the best linear fit.

As a necessary condition for a minimum of the expression (1.3) with respect to a and v, we have

$$\begin{aligned} \frac{\partial S}{\partial a} &= -2 \sum_{i=1}^{m} (y_i - a - vt_i) = 0 \\ \frac{\partial S}{\partial v} &= -2 \sum_{i=1}^{m} t_i (y_i - a - vt_i) = 0 \end{aligned} \tag{1.4}$$

This system of linear equations can be written in the following way:

$$\begin{aligned} \alpha_{11}\, a + \alpha_{12} v &= b_1 \\ \alpha_{21}\, a + \alpha_{22} v &= b_2 \end{aligned} \tag{1.5}$$

with

$$\alpha_{11} = m, \quad \alpha_{12} = \alpha_{21} = \sum_{i=1}^{m} t_i$$

$$\alpha_{22} = \sum_{i=1}^{m} t_i^2, \quad b_1 = \sum_{i=1}^{m} y_i, \quad b_2 = \sum_{i=1}^{m} t_i y_i \tag{1.6}$$

In matrix notation, equations (1.2) become

$$\mathbf{y} = \mathbf{A}\mathbf{x} + \mathbf{r}, \tag{1.7}$$

where

$$\mathbf{y} = \begin{bmatrix} y_1 \\ y_2 \\ \cdot \\ \cdot \\ \cdot \\ y_m \end{bmatrix}, \quad \mathbf{A} = \begin{bmatrix} 1 & t_1 \\ 1 & t_2 \\ \cdot & \\ \cdot & \\ \cdot & \\ 1 & t_m \end{bmatrix}, \quad \mathbf{x} = \begin{bmatrix} a \\ v \end{bmatrix}, \quad \mathbf{r} = \begin{bmatrix} r_1 \\ r_2 \\ \cdot \\ \cdot \\ \cdot \\ r_m \end{bmatrix}$$

Thus, the sum of the squares of the errors takes the form:

$$S = \mathbf{r}^T\mathbf{r} = (\mathbf{y} - \mathbf{A}\mathbf{x})^T(\mathbf{y} - \mathbf{A}\mathbf{x}) \tag{1.8}$$

By using this notation, we have, for (1.6):

$$\begin{bmatrix} \alpha_{11} & \alpha_{12} \\ \alpha_{21} & \alpha_{22} \end{bmatrix} = \mathbf{A}^T\mathbf{A}, \quad \begin{bmatrix} b_1 \\ b_2 \end{bmatrix} = \mathbf{A}^T\mathbf{y}$$

and for (1.5):

$$\mathbf{A}^T\mathbf{A}\mathbf{x} = \mathbf{A}^T\mathbf{y} \tag{1.9}$$

The solution of system (1.9) will be designated by $\hat{\boldsymbol{x}}$ and called the ***solution in the sense of the least-squares method*** of the overdetermined system of m equations with two unknowns:

$$\mathbf{A}\mathbf{x} = \mathbf{y} \tag{1.10}$$

Therefore, we have

$$\hat{\mathbf{x}} = (\mathbf{A}^T\mathbf{A})^{-1}\,\mathbf{A}^T\mathbf{y} \tag{1.11}$$

if $(\mathbf{A}^T\mathbf{A})^{-1}$ exists.

1.1.1 Example: Rolling ball

For table 1.1, we calculate $\mathbf{A}^T\mathbf{A}\hat{\mathbf{x}}=\mathbf{A}^T\mathbf{y}$:

$$\begin{bmatrix} 1 & 1 & 1 & 1 & 1 & 1 \\ 0 & 3 & 4 & 7 & 8 & 10 \end{bmatrix} \begin{bmatrix} 1 & 0 \\ 1 & 3 \\ 1 & 4 \\ 1 & 7 \\ 1 & 8 \\ 1 & 10 \end{bmatrix} \begin{bmatrix} \hat{a} \\ \hat{v} \end{bmatrix} = \begin{bmatrix} 1 & 1 & 1 & 1 & 1 & 1 \\ 0 & 3 & 4 & 7 & 8 & 10 \end{bmatrix} \begin{bmatrix} 1 \\ 4 \\ 5 \\ 6 \\ 7 \\ 10 \end{bmatrix}$$

After multiplying

$$\begin{bmatrix} 6 & 32 \\ 32 & 238 \end{bmatrix} \begin{bmatrix} \hat{a} \\ \hat{v} \end{bmatrix} = \begin{bmatrix} 33 \\ 230 \end{bmatrix}$$

from which we get

$$\hat{a}=1.22 \text{ and } \hat{v}=0.80$$

The corresponding straight line has been drawn in figure 1.2

1.1.2 Example: Navigation problem

In radio direction-finding navigation systems, we find the position of a ship by using a loop antenna to determine the bearings of at least two transmitters whose positions are known. The point of intersection of the lines thus obtained determines the position of the ship.

We will solve this problem by applying the least-squares method, and, as an example, we shall consider the case in which three transmitters are available.

To determine the position (x_0, y_0) of the ship in the xy-plane, we thus have three linear measurement equations.

$$\begin{aligned} a_1 x + b_1 y + c_1 &= r_1 \\ a_2 x + b_2 y + c_2 &= r_2 \\ a_3 x + b_3 y + c_3 &= r_3 \end{aligned}$$

in which the coefficents a_i, b_i and c_i are known. It is assumed that the three equations are normalized, i.e., that $a_i^2 + b_i^2 = 1$. In this case, the r_i terms represent the distances from the point (x_0, y_0) to the corresponding straight line (fig. 1.3).

By solving this problem according to the least-squares methods, we determine (x_0, y_0) such that the sum of these distances squared is a minimum.

Given, for example,

$$\begin{aligned} 0.6\,x_0 - 0.8\,y_0 + 1 &= r_1 \\ 0.8\,x_0 + 0.6\,y_0 + 2 &= r_2 \\ y_0 - 1 &= r_3 \end{aligned}$$

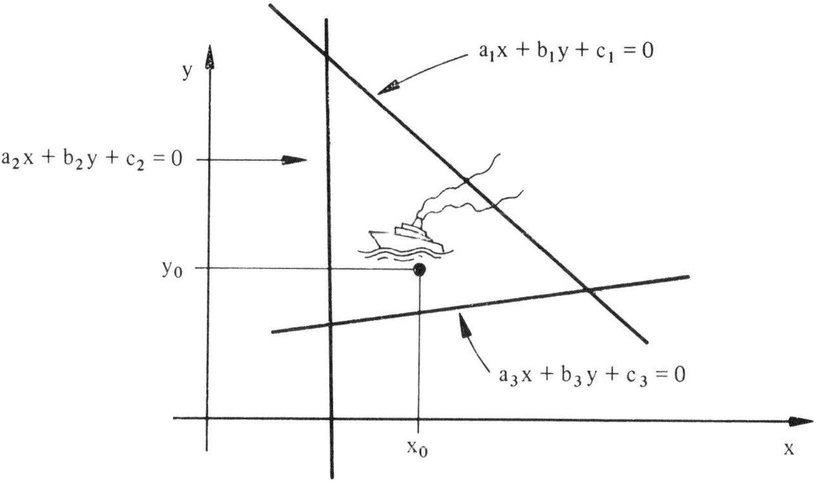

Fig. 1.3

In matrix notation, we have the measurement equation $\mathbf{y} = \mathbf{A}\mathbf{x} + \mathbf{r}$:

$$\begin{bmatrix} -1 \\ -2 \\ 1 \end{bmatrix} = \begin{bmatrix} 0.6 & -0.8 \\ 0.8 & 0.6 \\ 0 & 1 \end{bmatrix} \begin{bmatrix} x_0 \\ y_0 \end{bmatrix} + \begin{bmatrix} r_1 \\ r_2 \\ r_3 \end{bmatrix}$$

and for $\mathbf{A}^T\mathbf{A}\hat{\mathbf{x}} = \mathbf{A}^T\mathbf{y}$:

$$\begin{bmatrix} 1 & 0 \\ 0 & 2 \end{bmatrix} \begin{bmatrix} \hat{x}_0 \\ \hat{y}_0 \end{bmatrix} = \begin{bmatrix} -2.2 \\ 0.6 \end{bmatrix}$$

from which we get

$$\begin{bmatrix} \hat{x}_0 \\ \hat{y}_0 \end{bmatrix} = \begin{bmatrix} -2.2 \\ 0.3 \end{bmatrix}$$

1.1.3 Example: Trajectory of an airplane

Using the least-squares method, we attempt to determine the velocity vector $\mathbf{v}$ and initial position $\mathbf{x}_0$ of an airplane by means of a rotating surveillance radar. Assuming that its speed and direction of flight are constant, the plane's trajectory can be approximated by a straight line.

The positions of the airplane in an xyz-coordinate system are provided periodically by the radar with Δt being the scan period. Each component x, y, or z of the position vector is, therefore, a linear function of time. For the x-component, for example, we have

$$x_k = \xi_0 + k\,\Delta t v_x = r_k, \qquad k = 0,1,2\ldots, N$$

where the index k designates the *kth* measurement, ξ_0 is the x-coordinate at the starting time, v_x is the x-component of the velocity vector, and r_k is the measurement error.

In matrix notation, the measurment equation is presented as follows:

$$\underbrace{\begin{bmatrix} x_0 \\ x_1 \\ x_2 \\ \vdots \\ x_N \end{bmatrix}}_{\mathbf{y}} = \underbrace{\begin{bmatrix} 1 & 0 \\ 1 & 1 \\ 1 & 2 \\ \vdots & \\ 1 & N \end{bmatrix}}_{\mathbf{A}} \underbrace{\begin{bmatrix} \xi_0 \\ \Delta t v_x \end{bmatrix}}_{\mathbf{x}} + \underbrace{\begin{bmatrix} r_0 \\ r_1 \\ r_2 \\ \vdots \\ r_N \end{bmatrix}}_{\mathbf{r}}$$

According to formula (1.11), we calculate the estimate:

$$\begin{bmatrix} \hat{\xi}_0 \\ \Delta t \hat{v}_x \end{bmatrix} = (\mathbf{A}^T\mathbf{A})^{-1}\,\mathbf{A}^T\mathbf{y} = \frac{12}{N(N+2)} \begin{bmatrix} \frac{N(2N+1)}{6} & -\frac{N}{2} \\ -\frac{N}{2} & 1 \end{bmatrix} \begin{bmatrix} \overline{x} \\ \overline{M} \end{bmatrix}$$

where

$$\overline{x} = \frac{1}{N+1}\sum_{k=0}^{N} x_k, \qquad \overline{M} = \frac{1}{N+1}\sum_{k=1}^{N} k\,x_k$$

1.2 GENERAL METHOD

We generalize the problem for the case in which we wish to estimate n unknowns x_k on the basis of a number of measurements m that is larger than the number of unknowns ($m > n$).

In the case of linear measurements, the unknowns x_k must satisfy an overdetermined system of linear equations. In general, such a system has no solution unless measurement errors are assumed. Thus, we obtain the following system of linear equations:

$$
\begin{aligned}
a_{11} x_1 + a_{12} x_2 + \ldots + a_{1n}x_n &= y_1 + r_1 \\
a_{21} x_1 + a_{22} x_2 + \ldots + a_{2n}x_n &= y_2 + r_2 \\
&\vdots \\
a_{m1} x_1 + a_{m2} x_2 + \ldots + a_{mn}x_n &= y_m + r_m
\end{aligned}
$$

where the a_{jk} terms are the known coefficients of the measurement equations, y_k are the measurements, and r_k are the measurement errors.

In matrix notation, the system of m measurement equations has the form:

$$\mathbf{Ax} = \mathbf{y} + \mathbf{r} \tag{1.12}$$

where

$$\mathbf{A} = \begin{bmatrix} a_{11} & a_{12} \ldots & a_{1n} \\ a_{21} & a_{22} & a_{2n} \\ \vdots & & \\ a_{m1} & a_{m2} & a_{mn} \end{bmatrix}, \quad \mathbf{y} = \begin{bmatrix} y_1 \\ y_2 \\ \vdots \\ y_m \end{bmatrix}, \quad \mathbf{r} = \begin{bmatrix} r_1 \\ r_2 \\ \vdots \\ r_m \end{bmatrix}, \quad \mathbf{x} = \begin{bmatrix} x_1 \\ x_2 \\ \vdots \\ x_n \end{bmatrix}$$

The sum S of the squares of the measurement errors r_k equals

$$S = \mathbf{r}^T\mathbf{r} = (\mathbf{Ax} - \mathbf{y})^T(\mathbf{Ax} - \mathbf{y}) \tag{1.13}$$

and minimizing (1.13) with respect to x_k $(k = 1,2,\ldots, n)$ entails the necessary conditions:

$$\frac{\partial S}{\partial x_k} = 0, \quad k = 1,2,\ldots, n.$$

We can easily verify that these conditions are written in matrix notations as

$$\mathbf{A}^T\mathbf{A}\hat{\mathbf{x}} = \mathbf{A}^T\mathbf{y} \tag{1.14}$$

where $\hat{\mathbf{x}}$ provides the minimum of (1.13). Consequently, the ***least-squares solution*** $\hat{\mathbf{x}}$ is given by

$$\hat{\mathbf{x}} = (\mathbf{A}^T\mathbf{A})^{-1}\, \mathbf{A}^T\mathbf{y} \tag{1.15}$$

The inverse $(\mathbf{A}^T\mathbf{A})^{-1}$ exists if and only if the rank of **A** equals the number of columns of **A**. In addition, we can solve the system $\mathbf{y} = \mathbf{Ax}$ in a unique manner if **A** is a regular square matrix. In this case, we have

$$\mathbf{x} = \mathbf{A}^{-1}(\mathbf{A}^T)^{-1}\, \mathbf{A}^T y = (\mathbf{A}^T\mathbf{A})^{-1}\, \mathbf{A}^T\mathbf{y} = \hat{\mathbf{x}}$$

and clearly $S = \mathbf{r}^T\mathbf{r} = 0$.

Furthermore, it can be demonstrated that solution (1.15) furnishes an absolute minimum of (1.13).

1.2.1 Example: Multilevel terrain

While surveying a plot of land, we wish to estimate the altitudes x_1, x_2, and x_3 of three levels. This is done partly by measuring the altitudes directly and partly by measuring the differences in altitude among the three levels (fig. 1.4).

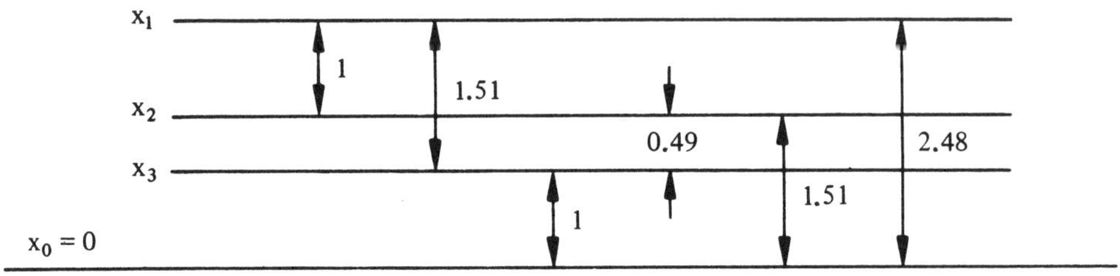

Fig. 1.4

We wish to estimate x_1, x_2, x_3 by the least-squares methods. The numerical values given determine the following system of measurement equations:

$$\mathbf{y} = \mathbf{A}\mathbf{x} + \mathbf{r}$$

where

$$\mathbf{A} = \begin{bmatrix} 1 & 0 & 0 \\ 0 & 1 & 0 \\ 0 & 0 & 1 \\ 1 & -1 & 0 \\ 0 & 1 & -1 \\ 1 & 0 & -1 \end{bmatrix}, \quad \mathbf{x} = \begin{bmatrix} x_1 \\ x_2 \\ x_3 \end{bmatrix}, \quad \mathbf{y} = \begin{bmatrix} 2.48 \\ 1.51 \\ 1 \\ 1 \\ 0.49 \\ 1.51 \end{bmatrix}, \quad \mathbf{r} = \begin{bmatrix} r_1 \\ r_2 \\ r_3 \\ r_4 \\ r_5 \\ r_6 \end{bmatrix}$$

Because the rank of $\mathbf{A} = 3$, $(\mathbf{A}^T\mathbf{A})^{-1}$ exists, and $\hat{\mathbf{x}}$ is given by $\hat{\mathbf{x}} = (\mathbf{A}^T\mathbf{A})^{-1}\mathbf{A}^T\mathbf{y}$. We calculate

$$\begin{bmatrix} \hat{x}_1 \\ \hat{x}_2 \\ \hat{x}_3 \end{bmatrix} = \begin{bmatrix} 3 & -1 & -1 \\ -1 & 3 & -1 \\ -1 & -1 & 3 \end{bmatrix}^{-1} \begin{bmatrix} 4.99 \\ 1 \\ -1 \end{bmatrix} = \begin{bmatrix} 2.50 \\ 1.50 \\ 1.00 \end{bmatrix}$$

1.3 WEIGHTED LEAST SQUARES

We consider the system $\mathbf{y} = \mathbf{A}\mathbf{x}$ and assume that the precision of the *kth* measurement is greater than the precision of the *ith* measurement. In this case, we shall give a greater "weight" to the *kth* equation than to the *ith* equation in the following sense: instead of minimizing $S = \mathbf{r}^T\mathbf{r}$ according to (1.13), we minimize

$$S = \sum_{i=1}^{m} \omega_i r_i^2 \tag{1.16}$$

where $\omega_i \geq 0$, $i = 1,2,\ldots,m$, are factors which determine the "weight" of the different equations.

It can be easily demonstrated that in this case, the ***solution*** $\hat{\mathbf{x}}$ in the sense of the least-squares method of $\mathbf{y} = \mathbf{A}\mathbf{x}$ is given by

$$\hat{\mathbf{x}} = (\mathbf{A}^T\mathbf{W}\mathbf{A})^{-1}\,\mathbf{A}^T\mathbf{W}\mathbf{y} \tag{1.17}$$

where

$$\mathbf{W} = \begin{bmatrix} \omega_1 & 0 & \ldots & 0 \\ 0 & \omega_2 & \ldots & 0 \\ \vdots & & & \\ 0 & & & \omega_m \end{bmatrix}$$

and assuming that $(\mathbf{A}^T\mathbf{W}\mathbf{A})^{-1}$ exists.

1.4 LEAST-SQUARES METHOD FOR NONLINEAR EQUATIONS

Let us assume that there are m different measurements of y: $y_1, y_2, \ldots, y_m$, and that the relationships between the y_i and n unknown parameters $x_i, x_2, \ldots, x_n$ are given by

$$y_i = f_i(x_1, x_2, \ldots, x_n), \quad i = 1,2,\ldots,m.$$

Instead of the system of linear equations (1.12), we have

$$\begin{aligned} f_1(x_1, x_2, \ldots, x_n) &= y_1 + r_1 \\ f_2(x_1, x_2, \ldots, x_n) &= y_2 + r_2 \\ &\vdots \\ f_m(x_1, x_2, \ldots, x_n) &= y_m + r_m \end{aligned} \tag{1.18}$$

where the r_i terms are, as before, measurement errors.

If we know an approximate solution $\mathbf{x}_0 = (x_{10}, x_{20}, \ldots, x_{n0})$, we can linearize system (1.18) by expanding the functions f_i, $i = 1,2,\ldots,m$, in Taylor series about the approximate solution. By retaining only the linear terms, we obtain

$$f_i(x_1, x_2, \ldots, x_n) \approx f_i(x_{10}, x_{20}, \ldots, x_{n0}) + \sum_{k=1}^{n} \frac{\partial f_i}{\partial x_k}(x_{10}, x_{20}, \ldots, x_{n0})\,\Delta x_k + r_i$$

where

$$x_k = x_{k0} + \Delta x_k, \quad k = 1,2\ldots, n$$

This means that instead of system (1.18), we consider the sytem of linear equations:

$$\mathbf{y} = \mathbf{A}\Delta\mathbf{x} + \mathbf{r}, \tag{1.19}$$

where

$$\mathbf{A} = \begin{bmatrix} \dfrac{\partial f_1}{\partial x_1} & \dfrac{\partial f_1}{\partial x_2} & \cdots & \dfrac{\partial f_1}{\partial x_n} \\ \dfrac{\partial f_2}{\partial x_1} & \dfrac{\partial f_2}{\partial x_2} & & \dfrac{\partial f_2}{\partial x_n} \\ \vdots & & & \\ \dfrac{\partial f_m}{\partial x_1} & \dfrac{\partial f_m}{\partial x_2} & & \dfrac{\partial f_m}{\partial x_n} \end{bmatrix}_{(x_{10}, x_{20}, \ldots, x_{n0})}, \quad \Delta\mathbf{x} = \begin{bmatrix} \Delta x_1 \\ \Delta x_2 \\ \vdots \\ \Delta x_n \end{bmatrix}$$

$$\mathbf{y} = \begin{bmatrix} y_1 - f_1(x_{10}, x_{20}, \ldots, x_{n0}) \\ y_2 - f_2(x_{10}, x_{20}, \ldots, x_{n0}) \\ \vdots \\ y_m - f_m(x_{10}, x_{20}, \ldots, x_{n0}) \end{bmatrix}, \quad \mathbf{r} = \begin{bmatrix} r_1 \\ r_2 \\ \vdots \\ r_m \end{bmatrix}$$

By solving this system according to the least-squares method, we have

$$\widehat{\Delta\mathbf{x}} = (\mathbf{A}^T\mathbf{A})^{-1}\,\mathbf{A}^T\mathbf{y} \tag{1.20}$$

If $(\mathbf{A}^T\mathbf{A})^{-1}$ exists.

The vector $\widehat{\Delta\mathbf{x}}$ is the estimate of the correction to which the approximate solution $\mathbf{x}_0$ is added to obtain a better estimate of the solution.

1.4.1 Example: Hyperbolic navigation

To determine the position P(x,y) of an airplane in an xy-coordinate system, we make use of three radio transmitters placed at A(0,0), B(4,0), and C(0,4), which periodically transmit pulsed signals at the same instant. By measuring the differences of the arrival times of the three signals, we obtain a direct measure of the three differences of the distances between the airplane and the three points A, B, and C:

$$|PA - PB| = 2, \quad |PA - PC| = 2, \quad |PB - PC| = 4$$

where PA, PB, and PC are the lengths of the segments $\overline{PA}$, $\overline{PB}$, and $\overline{PC}$.

Knowing the approximate position of the airplane P_0 (3.1; 0.1), we wish to estimate, with the aid of the method developed, the position correction Δx, Δy to obtain a more precise position estimate.

The solution is as follows: as $|PA - PB| = 2$, the x- and y- coordinates must satisfy the equation:

$$(x-2)^2 - \frac{y^2}{3} = 1 \qquad \text{(a)}$$

likewise, $|PB - PC| = 4$ implies that

$$(y-2)^2 - \frac{x^2}{3} = 1 \qquad \text{(b)}$$

and $|PB - PC| = 4$:

$$\frac{(x-y)^2}{8} - \frac{(x+y-4)^2}{8} = 1 \qquad \text{(c)}$$

The three loci expressed by relations (a), (b), and (c), therefore, are hyperbolas, from which the name of this method of navigation is derived (see fig. 1.5). By linearizing equations (a), (b), and (c) about the point P_0, we obtain the following system of linear equations:

$$(x_0-2)^2 + 2(x_0-2)\Delta x - \frac{y_0^2}{3} - \frac{2}{3}\, y_0 \Delta y = 1 \qquad \text{for (a)}$$

$$-\frac{x_0^2}{3} - \frac{2}{3}\, x_0 \Delta x + (y_0-2)^2 + 2(y_0-2)\Delta y = 1 \qquad \text{for (b)}$$

and

$$\frac{(x_0-y_0)^2}{8} + \frac{1}{4}(x_0-y_0)\Delta x - \frac{1}{4}(x_0-y_0)\Delta y - \frac{(x_0+y_0-4)^2}{8}$$

$$-\frac{1}{4}(x_0+y_0-4)\Delta x - \frac{1}{4}(x_0+y_0-4)\Delta y = 1 \qquad \text{for (c)}$$

By substituting the initial values $x_0 = 3.1$ and $y_0 = 0.1$, we have, in matrix form:

$$\begin{bmatrix} -0.20667 \\ 0.59333 \\ -0.045 \end{bmatrix} = \begin{bmatrix} 2.2 & -0.0666 \\ -2.0666 & -3.8 \\ 0.95 & -0.55 \end{bmatrix} \begin{bmatrix} \Delta x \\ \Delta y \end{bmatrix} + \begin{bmatrix} r_1 \\ r_2 \\ r_3 \end{bmatrix}$$

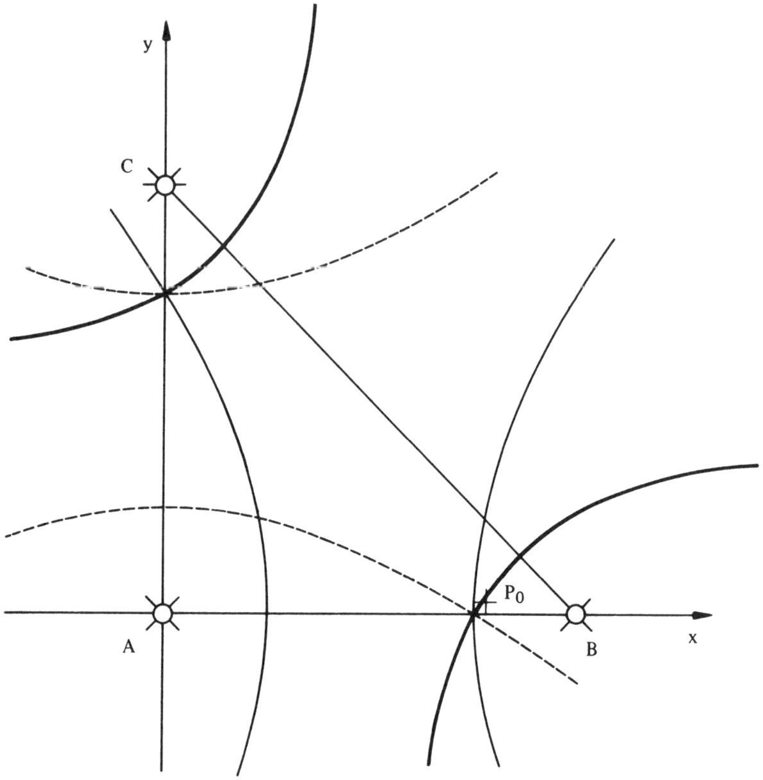

Fig. 1.5

From (1.20), we calculate

$$\begin{bmatrix} \widehat{\Delta x} \\ \widehat{\Delta y} \end{bmatrix} = (A^T A)^{-1} A^T y = \begin{bmatrix} -0.09887 \\ -0.10211 \end{bmatrix} \approx \begin{bmatrix} -0.1 \\ -0.1 \end{bmatrix}$$

Thus, P(3.1-0.1; 0.1-0.1) = P(3, 0) is a better estimate of the position of the airplane than P_0.

1.5 EXERCISES

1.5.1 Given a physical process described by the exponential function $y = A \exp(bt)$; after measuring y_k at the instants t_k, $k = 1,2,\ldots, n$, determine analytically the parameters ln A and b that minimize the sum of the squares of the errors $r_k = \ln y_k - \ln A - bt_k$.

1.5.2 By the same method, estimate the constant g in the law for a falling body $x = 1/2\ gt^2$, making use of the measurements of x_k at the instants t_k, $k = 1,2,\ldots,n$.

- estimate g by setting $x_k = 1/2\ gt_k^2 + r_k$;
- estimate g by setting $\ln x_k = \ln(g/2) + 2 \ln t_k + r_k$;
- compare the results.

1.5.3 Let us consider a discrete linear system whose law is given by the recursion formula $x_{k+1} = ax_k + b$. Estimate the parameters a and b by the least-squares method, given the measurements in table 1.6

Table 1.6

k	1	2	3	4	5
x_k	0.9	3.2	6.7	15.4	31

1.5.4 Estimate by the least-squares method the impact point of the parabolic trajectory that issues from the origin, with the following points obtained by means of a tracking radar: $P_1(1.4)$, $P_2(2.5)$, $P_3(3.6)$, and $P_4(4.3)$ (see fig. 1.7).

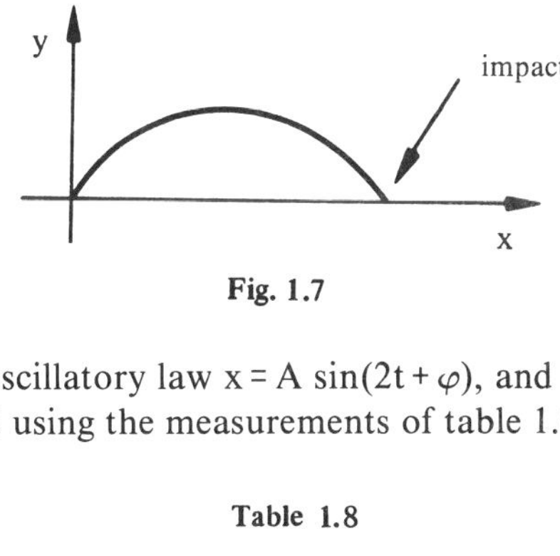

Fig. 1.7

1.5.5 Linearize the oscillatory law $x = A \sin(2t + \varphi)$, and estimate A and φ by the least-squares method using the measurements of table 1.8.

Table 1.8

t_k	0	$\pi/4$	$\pi/2$	$3\pi/4$
x_k	1.6	1.1	−1.8	−0.9

1.5.6 Estimate by the least-squares method the position (x,y) of a point P, after taking the following measurements of the tangents of the angles α, β, and γ: $\tan \alpha = 3/4$, $\tan \beta = 5/4$, $\tan \gamma = 5/4$ (see fig. 1.9).

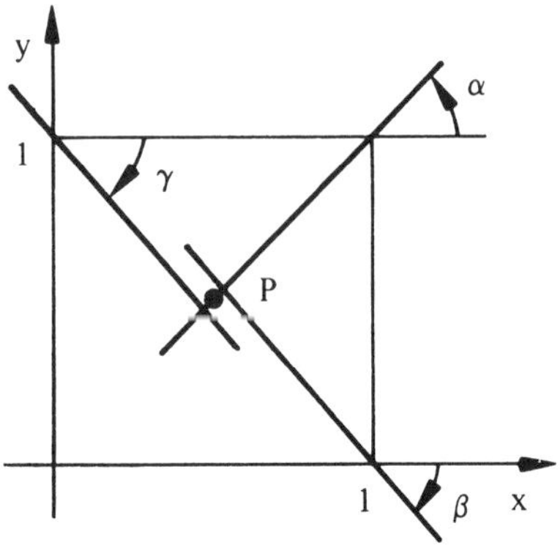

Fig. 1.9

1.5.7 Use the least-squares method to estimate the parameters a and b of the differential equation $\dot{x} = ax + bt^2$, by making the approximation:

$$\dot{x}_k \approx \frac{x_{k+1} - x_k}{\Delta t}$$

and given x(t) at the instants $t_k = 0,1,2,3$: $x(0) = 1$, $x(1) = 2$, $x(2) = -1$, $x(3) = 3$.

1.5.8 Find the "best" intersection of the three circles $x^2 + y^2 = 1$, $x^2 + y^2 - 2x = 0$, $(x+2)^2 + (y-2)^2 = 9$. First, obtain graphically an approximate point of intersection P_0 (x_0, y_0), then linearize the three equations about this point and estimate the correction Δx and Δy by the least-squares method.

1.5.9 In the case of the general method, demonstrate that the necessary conditions $\partial S/\partial x_k = 0$, $k = 1,2,\ldots,n$, to minimize the sum S of the squares of the errors allow the matrix representation $\mathbf{A}^T \mathbf{A}\hat{\mathbf{x}} = \mathbf{A}^T \mathbf{y}$ (1.14).

Chapter 2

Solution of Equations by Iterative Methods

2.1 INTRODUCTION

We shall consider the procedures of iterative calculations. To illustrate the basic idea of this type of method, we shall consider a very simple example, consisting of solving the equation:

$$x^2 - a = 0, \qquad a > 0 \tag{2.1}$$

In order to construct an iterative method for numerical calculation of the solution $\pm\sqrt{a}$ of (2.1), we shall consider the graph of the function $y(x) = x^2 - a$ (fig. 2.1).

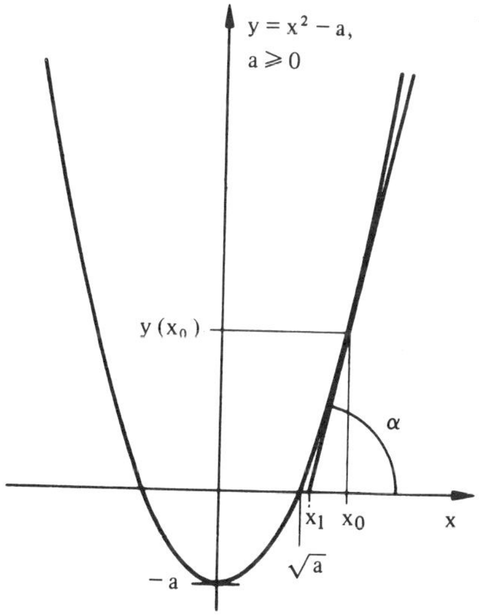

Fig. 2.1

To find the solution of equation (2.1), we must therefore go back to determining the points of intersection of the parabola $y = x^2 - a$ with the x – axis. Starting from an arbitrary point x_0, which we will choose near $\sqrt{a}$, we construct the tangent to the parabola at the point $(x_0, y(x_0))$. We find that the point of intersection x_1 of this tangent with the x – axis is generally closer to $\sqrt{a}$ than the starting point x_0. As

$$\tan \alpha = \frac{y(x_0)}{x_0 - x_1} = y'(x_0)$$

we have

$$x_1 = x_0 - \frac{y(x_0)}{y'(x_0)}$$

By repeating this procedure, we obtain a sequence (x_k) of points approaching $\sqrt{a}$:

$$x_{k+1} = x_k - \frac{y(x_k)}{y'(x_k)} = \frac{x_k}{2} + \frac{a}{2x_k}, \quad k = 0, 1, 2, \ldots$$

A calculation procedure of the form:

$$x_{k+1} = F(x_k), \quad k = 0, 1, 2, \ldots \tag{2.2}$$

is called an ***iterative method*** in which we start from a given value x_0 to calculate x_1, then with the help of x_1 we calculate x_2, *et cetera*. The formula itself is known as the ***recursion formula***. The procedure is called ***convergent*** if the sequence x_k approaches a finite number as k approaches $+\infty$, if not, it is called ***divergent***.

2.2 NEWTON-RAPHSON METHOD

The Newton-Raphson method is often used to solve equations of the form $f(x) = 0$.

We denote an exact root sought by x and an approximate value of x by x_0. Assuming that f(x) can be expanded in a Taylor series about x_0, we have

$$f(x) = f(x_0) + f'(x_0)(x - x_0) + \frac{f''(\xi)}{2}(x - x_0)^2$$

where, $\xi \in (x, x_0)$; and as $f(x) = 0$:

$$0 = f(x_0) + f'(x_0)(x - x_0) + \frac{f''(\xi)}{2}(x - x_0)^2$$

from which we calculate, if $f'(x_0) \neq 0$:

$$x = x_0 - \frac{f(x_0)}{f'(x_0)} - \frac{f''(\xi)}{2f'(x_0)}(x - x_0)^2 \qquad (2.3)$$

If we leave out the remainder:

$$R_2 = -\frac{f''(\xi)}{2f'(x_0)}(x - x_0)^2$$

then the right side of (2.3) no longer represents the exact root x, but an improved approximate root x_1 with respect to x_0. By repeating the procedure, we find the recurrence formula:

$$x_{k+1} = x_k - \frac{f(x_k)}{f'(x_k)}, \quad k = 0, 1, 2, \ldots \qquad (2.4)$$

which is called the ***Newton-Raphson*** recursion formula.

From a geometric point of view, the method consists of a linear extrapolation. Starting from a point x_k close to the root x, we extrapolate along the tangent in $(x_k, f(x_k))$ up to its point of intersection with the x-axis, and this point of intersection is called x_{k+1} (see fig. 2.2)

Since we have

$$\tan\alpha = \frac{f(x_k)}{\Delta x_k} = f'(x_k)$$

by setting $\Delta x_k = x_k - x_{k-1}$, again we find

$$x_{k+1} = x_k - \frac{f(x_k)}{f'(x_k)}, \qquad k = 0, 1, 2, \ldots$$

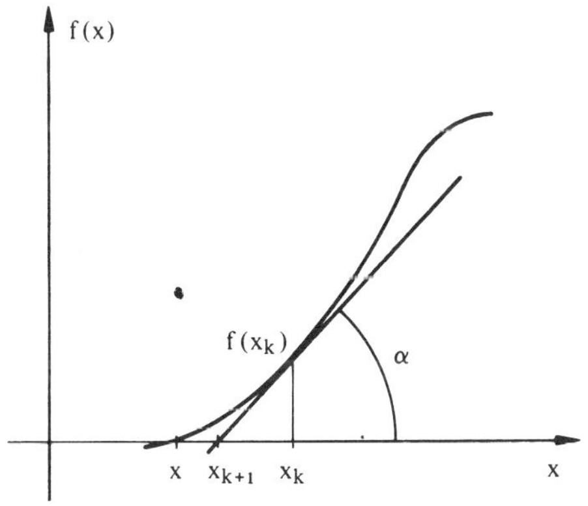

Fig. 2.2

It is clear that an iterative method is only useful if it converges toward the values sought. It is in this sense that we wish to study the convergence of the Newton-Raphson method in a more comprehensive manner at the end of this chapter (see sec. 2.4.5).

2.2.1 Example: Iterative calculation of the reciprocal value

Let us consider a calculator that can only carry out the operations of addition, subtraction, and multiplication. When $a > 0$ is given, we wish to calculate its reciprocal value $1/a$.

The problem can be reduced to solving the equation $f(x) = x^{-1} - a = 0$. According to the Newton-Raphson formula (2.4), we have

$$x_{k+1} = x_k - \frac{f(x_k)}{f'(x_k)} = 2x_k - a x_k^2$$

which is a recurrence that does not require division.

For $a = 7$ and starting from $x_0 = 0.2$, for example, we calculate

$$x_1 = 0.12 \qquad x_3 = 0.142\,763\,52$$

$$x_2 = 0.1392 \qquad x_4 = 0.142\,857\,081$$

This series converges toward $1/7 = 0.142857142\ldots$

2.2.2 Example: Comparison of different recursion formulas

We wish to solve the equation:

$$x = \exp(1/x) \tag{a}$$

By transforming the given equation in different ways, we arrive at different recursion formulas. For

$$f(x) = x - \exp(1/x)$$

the Newton-Raphson formula gives

$$x_{k+1} = x_k - x_k^2 \frac{x_k - \exp(1/x_k)}{x_k^2 + \exp(1/x_k)} \tag{b}$$

If we set $x = 1/y$, equation (a) becomes

$$1/y = \exp(y)$$

and for

$$f(y) = 1 - y \exp(y)$$

we find according to (2.4):

$$y_{k+1} = y_k + \frac{\exp(-y_k) - y_k}{1 + y_k} \tag{c}$$

By expressing equation (a) in logarithmic form, with $f(x) = 1 - x \ln x$, we obtain

$$x_{k+1} = x_k + \frac{1 - x_k \ln x_k}{1 + \ln x_k} \tag{d}$$

Thus, we obtain three different recursion formulas for the same problem. It is clear that formulas (c) and (d) are better suited to calculation than formula (b).

The graphical representation of $f(x) = x - \exp(1/x)$ shows that there is only a single root x, and it must be found close to $x_0 = 1.8$.

Starting from $y_0 = 0.6 \approx 1/1.8$, we calculate according to (c):

$y_1 = 0.568\,007$ $\quad y_3 = 0.567\,143,$ *et cetera*
$y_2 = 0.567\,144$ $\quad x \approx 1/y_3 = 1.763\,223$

and, according to (d), starting from $x_0 = 1.8$:

$x_1 = 1.763\,461$ $\quad x_3 = 1.763\,223,$ *et cetera*
$x_2 = 1.763\,223$ $\quad x \approx 1.763\,223$

2.3 NEWTON-RAPHSON METHOD FOR TWO UNKNOWNS

We wish to solve the nonlinear system of two equations:

$$F(x, y) = 0 \qquad G(x, y) = 0 \tag{2.5}$$

where F and G are given functions of the independent variables x and y.

In this case, the Newton-Raphson method is analogous to that which was presented in section 2.2: after choosing an approximate solution (x_0, y_0), i.e., an initial approximation, we set

$$x = x_0 + \epsilon_0$$
$$y = y_0 + \tau_0$$

and expand the two functions F and G in Taylor series about the point (x_0, y_0). Since $(x_0 + \epsilon, y_0 + \tau_0)$ is a solution of the system of equations (2.5), the results are

$$0 = F(x_0 + \epsilon_0, y_0 + \tau_0) = F(x_0, y_0) + \frac{\partial F}{\partial x}\epsilon_0 + \frac{\partial F}{\partial y}\tau_0 + \text{higher order terms}$$

$$0 = G(x_0 + \epsilon_0, y_0 + \tau_0) = G(x_0, y_0) + \frac{\partial G}{\partial x}\epsilon_0 + \frac{\partial G}{\partial y}\tau_0 + \text{higher order terms}$$

or, in matrix form:

$$\begin{bmatrix} 0 \\ 0 \end{bmatrix} = \begin{bmatrix} F \\ G \end{bmatrix}_{(x_0,y_0)} + \begin{bmatrix} \frac{\partial F}{\partial x} & \frac{\partial F}{\partial y} \\ \frac{\partial G}{\partial x} & \frac{\partial G}{\partial y} \end{bmatrix}_{(x_0,y_0)} \begin{bmatrix} \epsilon_0 \\ \tau_0 \end{bmatrix} + \text{higher order terms}$$

Assuming the existence of

$$\begin{bmatrix} \frac{\partial F}{\partial x} & \frac{\partial F}{\partial y} \\ \frac{\partial G}{\partial x} & \frac{\partial G}{\partial y} \end{bmatrix}^{-1}$$

we have

$$\begin{bmatrix} \epsilon_0 \\ \tau_0 \end{bmatrix} = \begin{bmatrix} x \\ y \end{bmatrix} - \begin{bmatrix} x_0 \\ y_0 \end{bmatrix} = - \begin{bmatrix} \frac{\partial F}{\partial x} & \frac{\partial F}{\partial y} \\ \frac{\partial G}{\partial x} & \frac{\partial G}{\partial y} \end{bmatrix}^{-1}_{(x_0,y_0)} \begin{bmatrix} F \\ G \end{bmatrix}_{(x_0,y_0)} + \text{higher order terms}$$

which gives as a solution (x,y):

$$\begin{bmatrix} x \\ y \end{bmatrix} = \begin{bmatrix} x_0 \\ y_0 \end{bmatrix} - \begin{bmatrix} \dfrac{\partial F}{\partial x} & \dfrac{\partial F}{\partial y} \\ \dfrac{\partial G}{\partial x} & \dfrac{\partial G}{\partial y} \end{bmatrix}^{-1}_{(x_0,y_0)} \begin{bmatrix} F \\ G \end{bmatrix}_{(x_0,y_0)} + \begin{matrix}\text{higher} \\ \text{order} \\ \text{terms}\end{matrix} \qquad (2.6)$$

When we neglect the higher-order terms, the solution (x,y) becomes an approximation, which will be taken as the new starting point (x_1, y_1) for the next approximation:

$$\begin{bmatrix} x_1 \\ y_1 \end{bmatrix} = \begin{bmatrix} x_0 \\ y_0 \end{bmatrix} - \begin{bmatrix} \dfrac{\partial F}{\partial x} & \dfrac{\partial F}{\partial y} \\ \dfrac{\partial G}{\partial x} & \dfrac{\partial G}{\partial y} \end{bmatrix}^{-1}_{(x_0,y_0)} \begin{bmatrix} F \\ G \end{bmatrix}_{(x_0,y_0)}$$

Thus, we find the recursion formula of ***Newton-Raphson*** for two unknowns.

$$\begin{bmatrix} x_{k+1} \\ y_{k+1} \end{bmatrix} = \begin{bmatrix} x_k \\ y_k \end{bmatrix} - \begin{bmatrix} \dfrac{\partial F}{\partial x} & \dfrac{\partial F}{\partial y} \\ \dfrac{\partial G}{\partial x} & \dfrac{\partial G}{\partial y} \end{bmatrix}^{-1}_{(x_k,y_k)} \begin{bmatrix} F \\ G \end{bmatrix}_{(x_k,y_k)} , \quad k = 0, 1, 2, \ldots \qquad (2.7)$$

2.3.1 Example: Tranformation into a system with two unknowns

We wish to solve the equation:

$$x = 2 \sin x \qquad \text{(a)}$$

in two different ways.

On the one hand, we set $f(x) = x - 2\sin x = 0$, which gives, using formula (2.4), for one variable:

$$x_{k+1} = x_k - \frac{f(x_k)}{f'(x_k)} = x_k - \frac{x - 2\sin x_k}{1 - 2\cos x_k}$$

On the other hand, we factor (a) into a system of two equations:

$$y = x$$
$$y = 2 \sin x$$

where, in order to apply (2.7),

$$F = y - x = 0$$
$$G = y - 2 \sin x = 0$$

Thus, we obtain

$$\begin{bmatrix} x_{k+1} \\ y_{k+1} \end{bmatrix} = \begin{bmatrix} x_k \\ y_k \end{bmatrix} - \begin{bmatrix} -1 & 1 \\ -2\cos x_k & 1 \end{bmatrix}^{-1} \begin{bmatrix} y_k - x_k \\ y_k - 2\sin x_k \end{bmatrix}$$

The starting point can be chosen with the aid of a graph, for example, of $x_0 = 2$, $y_0 = 1.8$.

2.3.2 Example: Maxima and minima of a function of two variables

We wish to determine the maxima and minima of a function f(x,y).

Assuming that f(x,y) is sufficiently differentiable, we expand f(x,y) in a Taylor series about the point (x,y) at which the function has an extremum:

$$f(x+\Delta x, y+\Delta y) = f(x,y) + \frac{\partial f}{\partial x}\Delta x + \frac{\partial f}{\partial y}\Delta y + \frac{1}{2}\left\{\frac{\partial^2 f}{\partial x^2}\Delta x^2 + \right.$$
$$\left. + 2\frac{\partial^2 f}{\partial x \partial y}\Delta x \Delta y + \frac{\partial^2 f}{\partial y^2}\Delta y^2\right\} + \text{higher-order terms}$$

In order for f(x,y) to have an extremum at (x,y), it is necessary that $\partial f/\partial x = \partial f/\partial y = 0$. Thus, leaving out the terms of order ≥ 3, we find

$$\Delta f = f(x+\Delta x, y+\Delta y) - f(x,y)$$
$$= \frac{1}{2}\left\{\frac{\partial^2 f}{\partial x^2}\Delta x^2 + 2\frac{\partial^2 f}{\partial x \partial y}\Delta x \Delta y + \frac{\partial^2 f}{\partial y^2}\Delta y^2\right\}$$

or, in matrix form:

$$\Delta f = \frac{1}{2}\begin{bmatrix}\Delta x & \Delta y\end{bmatrix}\begin{bmatrix}\dfrac{\partial^2 f}{\partial x^2} & \dfrac{\partial^2 f}{\partial x \partial y} \\ \dfrac{\partial^2 f}{\partial x \partial y} & \dfrac{\partial^2 f}{\partial y^2}\end{bmatrix}_{(x,y)}\begin{bmatrix}\Delta x \\ \Delta y\end{bmatrix}$$

The matrix

$$\mathbf{S} = \begin{bmatrix}\dfrac{\partial^2 f}{\partial x^2} & \dfrac{\partial^2 f}{\partial x \partial y} \\ \dfrac{\partial^2 f}{\partial x \partial y} & \dfrac{\partial^2 f}{\partial y^2}\end{bmatrix}$$

is symmetrical, i.e., that $S^T = S$, where S^T is the transpose of S. Thus, its eigenvalues λ_1 and λ_2 are real, and there exists an orthogonal matrix P, such that

$$PSP^T = \begin{bmatrix}\lambda_1 & 0 \\ 0 & \lambda_2\end{bmatrix}$$

(see also section 4.2). Thus, we have

$$\Delta f = \frac{1}{2}\begin{bmatrix}\Delta x & \Delta y\end{bmatrix} P^T \begin{bmatrix}\lambda_1 & 0 \\ 0 & \lambda_2\end{bmatrix} P \begin{bmatrix}\Delta x \\ \Delta y\end{bmatrix} = \frac{1}{2}(\Delta\bar{x}^2 \lambda_1 + \Delta\bar{y}^2 \lambda_2)$$

where

$$\begin{bmatrix}\Delta\bar{x} \\ \Delta\bar{y}\end{bmatrix} = P\begin{bmatrix}\Delta x \\ \Delta y\end{bmatrix}$$

The signs of λ_1 and λ_2 then provide us with the information sought:

$\lambda_1, \lambda_2 > 0$	$\Rightarrow$	$\Delta f > 0$: f has a minimum at (x,y)
$\lambda_1, \lambda_2 < 0$	$\Rightarrow$	$\Delta f < 0$: f has a maximum at (x,y)
sign $\lambda_1 \neq$ sign λ_2	$\Rightarrow$	$\Delta f \lesseqgtr 0$: f has a saddle point at (x,y)
$\lambda_1 = 0, \lambda_2 \neq 0$	$\Rightarrow$	$\Delta f \geq 0$: f has a trough at (x,y) $\Delta f \leq 0$: f has an inverted trough at (x,y)

$\lambda_1 = \lambda_2 = 0 \quad \Rightarrow$ examine the higher-order terms in the Taylor series

We shall consider the numerical example:

$$f(x, y) = x^2 + 2xy + 2y^2 - 3x - 4y$$

Setting

$$\partial f/\partial x = 2x + 2y - 3 = 0$$
$$\partial f/\partial y = 2x + 4y - 4 = 0$$

we see that f has an extremum at (1;0.5). At this point, we calculate

$$S = \begin{bmatrix} 2 & 2 \\ 2 & 4 \end{bmatrix}$$

whose eigenvalues are $\lambda_1 = 5.236$ and $\lambda_2 = 0.764 > 0$, i.e., f has a minimum at (1;0.5).

Up to this point, we have proceeded in a purely analytical manner. If it is necessary to use numerical methods, especially in the case of more than two variables, we can, for example, apply the Newton-Raphson method to determine the points at which f has an extremum (for calculating the characteristic values of S see sec. 4.3 and 4.4).
For this numerical example, we will, therefore, solve the system of two equations:

$$F(x, y) = \partial f/\partial x = 2x + 2y - 3 = 0$$
$$G(x, y) = \partial f/\partial y = 2x + 4y - 4 = 0$$

using the recurrence formula (2.7), which gives

$$\begin{bmatrix} x_{k+1} \\ y_{k+1} \end{bmatrix} = \begin{bmatrix} x_k \\ y_k \end{bmatrix} - \frac{1}{4}\begin{bmatrix} 4 & -2 \\ -2 & 2 \end{bmatrix}\begin{bmatrix} 2x_k + 2y_k - 3 \\ 2x_k + 4y_k - 4 \end{bmatrix} = \begin{bmatrix} x_k \\ y_k \end{bmatrix} - \begin{bmatrix} x_k - 1 \\ y_k - 0.5 \end{bmatrix}$$

By choosing (0.5;1) as the starting point (x_0, y_0), we calculate

$$\begin{bmatrix} x_1 \\ y_1 \end{bmatrix} = \begin{bmatrix} 0.5 \\ 1 \end{bmatrix} - \begin{bmatrix} 0.5-1 \\ 1-0.5 \end{bmatrix} = \begin{bmatrix} 1 \\ 0.5 \end{bmatrix}$$

$$\begin{bmatrix} x_2 \\ y_2 \end{bmatrix} = \begin{bmatrix} 1 \\ 0.5 \end{bmatrix} - \begin{bmatrix} 0 \\ 0 \end{bmatrix} = \begin{bmatrix} 1 \\ 0{,}5 \end{bmatrix}$$

We have established that only a single step is necessary to find the exact solution. This results from the fact that $\partial f/\partial x$ and $\partial f/\partial y$ are linear functions. The terms of order greater than or equal to two cancel out in formula (2.6). Thus, the recursion formula (2.7) provides the exact solution after a single iteration.

2.4 FIXED-POINT METHOD

The solution of a problem using a recurrence formula $x_{k+1} = f(x_k)$, $k = 0,1,2,\ldots$, can be considered as the determination of a ***fixed point*** of the function f, i.e., of a point x^*, such that

$$x^* = f(x^*)$$

2.4.1 Fixed-point theorem

Let us assume that f, defined on the closed interval $\mathbf{I} = [a,b]$, satifies the following conditions

- condition *i*: $a \leq f(x) \leq b$ for all $x \in \mathbf{I}$
- condition *ii*: there exists a constant L, $0 \leq L < 1$, such that $|f(x) - f(y)| \leq L\,|x - y|$, where $x,y \in \mathbf{I}$

Under these conditions, f has on **I** exactly one fixed point $x^* = f(x^*)$, where x^* is the limit of the sequence,

$$x_{k+1} = f(x_k), \qquad k = 0, 1, 2, \ldots \tag{2.8}$$

and does not depend on the choice of the starting point $x_0 \in \mathbf{I}$.

We remark that condition *ii*, on the one hand, implies that f(x) is continuous on **I** and, on the other hand, by dividing the inequality by $x - y$ and passing to the limit by letting y tend toward x, the inequality $|f'(x)| \leq L < 1$ is obtained, assuming that the limit exists.

2.4.2 Demonstration of the fixed-point theorem

Let us first verify the ***convergence*** of the sequence (x_k). According to condition *ii*, we have

$$|x_{k+1} - x_k| \leqslant L|x_k - x_{k-1}| \leqslant \ldots \leqslant L^k |x_1 - x_0|$$

and for $n > k$, setting $x_n - x_k = x_n - x_{n-1} + x_{n-1} - x_{n-2} + \ldots - x_k$, and then applying the triangular inequality:

$$\begin{aligned} |x_n - x_k| &\leqslant |x_n - x_{n-1}| + \ldots + |x_{k+1} - x_k| \\ &\leqslant (L^{n-1} + \ldots + L^k)|x_1 - x_0| \\ &= L^k (L^{n-1-k} + \ldots + 1)|x_1 - x_0| \\ &\leqslant \frac{L^k}{1-L}|x_1 - x_0| \end{aligned}$$

where we have used the formula $1 + L + L^2 + \ldots = 1/(1-L)$. For each $\epsilon > 0$ there is, therefore, a value of k, such that $|x_n - x_k| < \epsilon$ for all $n > k$, i.e., that (x_k) is a Cauchy sequence. It converges toward a limit x^* belonging to the interval **I**, as a result of condition *i*.

Let us show that x^* is a ***fixed point***. The recurrence formula $x_{k+1} = f(x_k)$ and the triangular inequality imply the inequality:

$$|f(x^*) - x^*| \leqslant |f(x^*) - f(x_k)| + |x_{k+1} - x^*|$$

Because x_k converges to x^* as k approaches $+\infty$, and since f is continuous, we find that $f(x^*) = x^*$.

We must still demonstrate that x^* is the ***only fixed point***. Let us assume that there is another fixed point y^*. We have $|y^* - x^*| = f(y^*) - f(x^*)| \leq L|y^* - x^*|$, and since $L < 1$, we find that $y^* = x^*$.

2.4.3 Geometric interpretation of the fixed-point theorem

Figures 2.3 and 2.4 give the geometric interpretation of the fixed-point theorem.

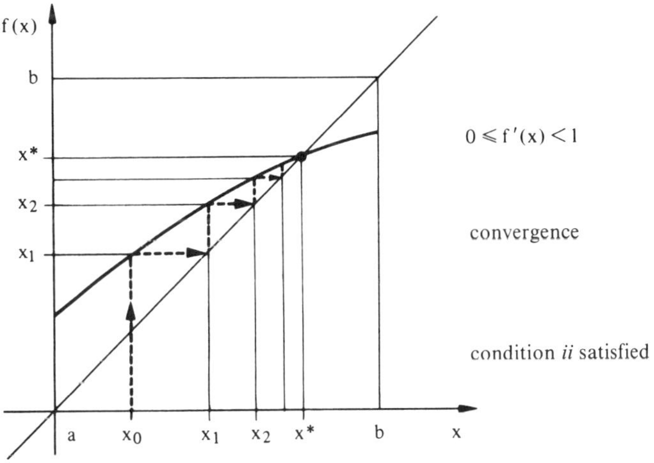

Fig. 2.3

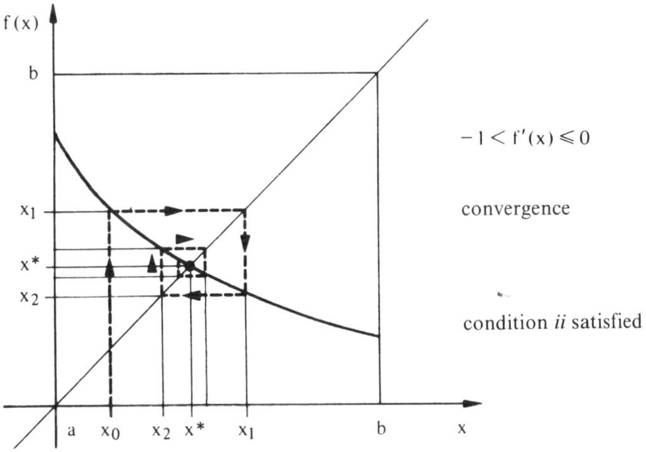

Fig. 2.4

If $|f'(x)| > 1$, condition *ii* is not satisfied. In this case, the recurrence can be divergent. This situation is illustrated by figure 2.5.

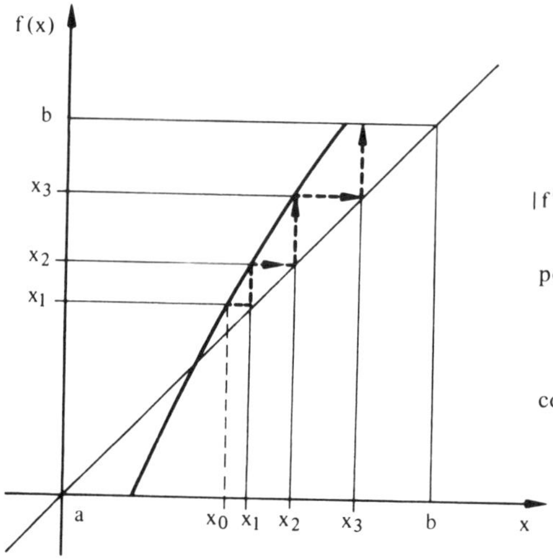

|f'(x)|> 1

possible divergence

condition *ii* not satisfied

Fig. 2.5

2.4.4 Example: Iterative resolution of the quadratic equation

We wish to solve the equation:

$$x^2 - 100x + 1 = 0 \qquad \text{(a)}$$

using the fixed-point method.

By writing (a) in the form:

$$x = \frac{1}{100}(x^2 + 1)$$

we obtain the recurrence formula:

$$x_{k+1} = f(x_k) = \frac{1}{100}(x_k^2 + 1) \qquad \text{(b)}$$

Because $f'(x) = 0.02x$, we can expect convergence of the sequence (b), if we start from a small value of x_0, for example, from $x_0 = 1$. We calculate

$$x_1 = 0.02 \qquad x_3 = 0.010\,001$$

$$x_2 = 0.010\,004 \qquad x_4 = 0.010\,001 \textit{ et cetera}$$

Thus, $x^* \approx 0.01$.

To determine the other root of equation (a), we shall write (a) in the form:

$$x^2 = 100x - 1 \tag{c}$$

and because this root will be different from 0, we can divide equation (c) by x. Thus we find the recurrence formula:

$$x_{k+1} = f(x_k) = 100 - \frac{1}{x_k} \tag{d}$$

Because $f'(x) = 1/x^2$, we can expect convergence of the sequence (d), if we start from a large value of x_0, for example, from $x_0 = 10$. We calculate

$$x_1 = 99.9 \qquad x_3 = 99.989\,999$$

$$x_2 = 99.989\,99 \qquad x_4 = 99.989\,999 \text{ et cetera}$$

Thus, $x^* \approx 99.99$.

2.4.5 Example: Convergence of the Newton-Raphson method

Let us again consider the Newton-Raphson formula (2.4)

$$x_{k+1} = x_k - \frac{f(x_k)}{f'(x_k)}, \qquad k = 0, 1, 2, \ldots$$

Thus, we have $x_{k+1} = g(x_k)$, where

$$g(x_k) = x_k - \frac{f(x_k)}{f'(x_k)}$$

If $|g'(x)| < 1$ for all the values of x close to a root x^* of the equation $f(x) = 0$, conditions *i* and *ii* of the fixed-point theorem 2.4.1 are satisfied. Thus, we have convergence of the sequence (x_k) toward a root x^* of the equation $f(x) = 0$.

By searching for x^* in the interval [a,b], we first note that the inequality $|g'(x)| < 1$ implies

$$|g'(x)| = \left| \left\{ x - \frac{f(x)}{f'(x)} \right\}' \right| = \left| \frac{f(x) f''(x)}{(f'(x))^2} \right| < 1$$

Let us assume that for all the x values of the interval [a,b] we have $|f'(x)| \geq m > 0$ and $|f''(x)| < M$. By developing f(x) around the approximate solution x_k, we have

$$f(x) = f(x_k) + f'(x_k)(x - x_k) + \frac{1}{2} f''(\xi_k)(x - x_k)^2$$

where $\xi_k \in (x, x_k)$. Because $f(x) = 0$ for $x = x^*$, we find

$$0 = f(x_k) + f'(x_k)(x^* - x_k) + \frac{1}{2} f''(\xi_k)(x^* - x_k)^2$$

and, dividing this equation by $f'(x_k) \neq 0$:

$$-\frac{f(x_k)}{f'(x_k)} = (x^* - x_k) + \frac{\frac{1}{2} f''(\xi_k)(x^* - x_k)^2}{f'(x_k)}$$

According to the Newton-Raphson formula (2.4), we have

$-f(x_k)/f'(x_k) = x_{k+1} - x_k$

Thus, we arrive at

$$x_{k+1} - x^* = \frac{\frac{1}{2} f''(\xi_k)}{f'(x_k)} (x_k - x^*)^2$$

and because $|f''(x)| \leq M$ and $|f'(x)| \geq m > 0$, we find the inequality:

$$|x_{k+1} - x^*| \leqslant \frac{M}{2m} |x_k - x^*|^2 \tag{a}$$

It has been established that the error of the (k + 1)th approximation is proportional to the square of the error of the *k*th approximation. For this reason, we say that the Newton-Raphson is a ***second-order recursive method***.

When we set $C = M/2m$, inequality (a) implies the following inequalities:

$$\begin{aligned} |x_{k+1} - x^*| &\leqslant C|x_k - x^*|^2 \leqslant C^3|x_{k-1} - x^*|^4 \leqslant C^7|x_{k-2} - x^*|^8 \leqslant \ldots \\ &\leqslant C^{2^{k+1}-1}|x_0 - x^*|^{2^{k+1}} \end{aligned}$$

Because $|x_0 - x^*| \leq |b - a|$, we arrive at the inequality:

$$|x_k - x^*| \leqslant C^{2^k - 1}|b - a|^{2^k} = \frac{[C(b-a)]^{2^k}}{C}, \quad k = 0, 1, 2, \ldots$$

Therefore, there is convergence toward the solution x^*, if $(M/2m)(b-a)<1$, which we can always arrange by choosing the interval $[a,b]$ sufficiently small, i.e., such that $(b-a)<2m/M$.

2.4.6 Example: Picard's method for the resolution of differential equations

Given the first-order differential equation:

$$y' = F(x,y), \qquad y(x_0) = y_0 \tag{a}$$

We seek to find a solution of (a), i.e., a function $y = y(x)$, such that $y'(x) = F(x,y(x))$ and $y(x_0) = y_0$.

If $F(x,y)$ is continuous, the function:

$$h(x) = y_0 + \int_{x_0}^{x} F(x, y(x))\,dx \tag{b}$$

exists for each continuous function $y(x)$. For $y = y(x)$ to be a solution of (a), it is necessary and sufficient that

$$y(x) = y_0 + \int_{x_0}^{x} F(x, y(x))\,dx$$

Solving (a) thus becomes a question of determining a "fixed point" of (b), i.e., a function $y(x)$, such that $h(x) = y(x)$.

Starting with the constant function $y_0(x) \equiv y_0$, we can always examine if the sequence $(y_k(x))$ of the functions:

$$y_{k+1}(x) = y_0 + \int_{x_0}^{x} F(x, y_k(x))\,dx, \quad k = 0, 1, 2, \ldots$$

converges toward a "fixed point." Because this method requires the evaluation of an infinite number of indefinite integrals, it does not generally lend itself to numerical calculation.

Applying this method to the solution of the equation:

$$y' = y + 1, \; y(0) = 1 \tag{c}$$

we find, starting from $y_0 = 1$:

$$y_{k+1} = 1 + \int_0^x (y_k + 1)\,dx$$

thus,

$$y_1 = 1 + 2x$$

$$y_2 = 1 + 2\left(x + \frac{x^2}{2!}\right)$$

$$y_3 = 1 + 2\left(x + \frac{x^2}{2!} + \frac{x^3}{3!}\right)$$

$$\vdots$$

$$y_k = 1 + 2\left(x + \frac{x^2}{2!} + \frac{x^3}{3!} + \ldots + \frac{x^k}{k!}\right)$$

$$\vdots$$

The sequence (y_k) converges toward a "fixed point," and we obtain the following as a solution of (c):

$$y(x) = \lim_{k \to \infty} y_k = 1 + 2(e^x - 1) = 2e^x - 1$$

2.5 JACOBI'S ALGORITHM

Jacobi's algorithm is an iterative method for solving a system of linear equations of the form:

$$\mathbf{A}\mathbf{x} = \mathbf{b} \tag{2.9}$$

where the diagonal elements of the coefficient matrix **A** are ***dominant*** with

$$A = \begin{bmatrix} a_{11} & a_{12} \ldots a_{1n} \\ a_{21} & \\ \vdots & \\ a_{n1} & a_{nn} \end{bmatrix}, \quad \mathbf{x} = \begin{bmatrix} x_1 \\ x_2 \\ \vdots \\ x_n \end{bmatrix} \quad \mathbf{b} = \begin{bmatrix} b_1 \\ b_2 \\ \vdots \\ b_n \end{bmatrix}$$

Since the diagonal elements $a_{ii} \neq 0$, we divide the *i*th equation, $i = 1,2,\ldots,n$, by its diagonal element a_{ii}, and solve the resulting equation for the unknown x_i. Thus, system (2.9) is transformed into

$$\mathbf{x} = -\mathbf{A}^*\mathbf{x} + \mathbf{b}^* \tag{2.10}$$

where

$$\mathbf{A}^* = \begin{bmatrix} 0 & \frac{a_{12}}{a_{11}} & \frac{a_{13}}{a_{11}} & \cdots & \frac{a_{1n}}{a_{11}} \\ \frac{a_{21}}{a_{22}} & 0 & \frac{a_{23}}{a_{22}} & & \frac{a_{2n}}{a_{22}} \\ \frac{a_{31}}{a_{33}} & & & & \\ \vdots & & & & \\ \frac{a_{n1}}{a_{nn}} & & & & 0 \end{bmatrix}, \quad \mathbf{b}^* = \begin{bmatrix} \frac{b_1}{a_{11}} \\ \frac{b_2}{a_{22}} \\ \frac{b_3}{a_{33}} \\ \vdots \\ \frac{b_n}{a_{nn}} \end{bmatrix}$$

We have thus reduced the problem of solving system (2.9) to the determination of a fixed point of the vector function:

$$\mathbf{f}(\mathbf{x}) = -\mathbf{A}^*\mathbf{x} + \mathbf{b}^*$$

The fixed-point recursion formula is given by (2.8):

$$\mathbf{x}_{k+1} = -\mathbf{A}^*\mathbf{x}_k + \mathbf{b}^*, \qquad k = 0, 1, 2, \ldots \tag{2.11}$$

Assuming that the diagonal elements are dominant, i.e., that

$$|a_{ii}| > \sum_{k \neq i, k=1}^{n} |a_{ik}|, \qquad i = 1, 2, \ldots, n$$

we demonstrate that $\mathbf{x}_k$ converges to the unique solution $\mathbf{x}$, starting from any initial vector $\mathbf{x}_0$.

To clarify the notation used in the demonstration, we shall write $\mathbf{x}^{(k)}$ instead of $\mathbf{x}_k$, the subscript still being reserved to designate the *k*th component of a vector. If the diagonal elements are dominant, the inverse $\mathbf{A}^{-1}$ of $\mathbf{A}$ exists (we state this fact without demonstration), and system (2.9) has exactly one solution $\mathbf{x}$. Therefore, we can set $\mathbf{x}^{(k)} = \mathbf{x} + \epsilon^{(k)}$, where $\mathbf{x}^{(k)}$ is determined according to (2.11) and where $\epsilon^{(k)}$ designates the error vector of the approximation $\mathbf{x}^{(k)}$. We have

$$x^{(k+1)} = x + \epsilon^{(k+1)} = -A^*(x + \epsilon^{(k)}) + b^*$$

and since $\mathbf{x} = -\mathbf{A}^* \mathbf{x} + \mathbf{b}^*$ (see (2.10)):

$$\epsilon^{(k+1)} = -A^* \epsilon^{(k)}$$

Let us consider the matrix $-\mathbf{A}^* = [a^*_{ij}]$. Because the diagonal elements are dominant, there is a constant C, $0 < C < 1$, such that

$$\sum_{j \neq i,\, j=1}^{n} |a^*_{ij}| \leqslant C < 1, \qquad i = 1, 2, ..., n$$

Consider $\epsilon_j^{(k+1)}$, which is the *j*th component of the error vector $\epsilon^{(k+1)}$. Because

$$\epsilon_j^{(k+1)} = \sum_{i=1}^{n} a^*_{ji}\, \epsilon_i^{(k)}$$

we have

$$|\epsilon_j^{(k+1)}| \leqslant \sum_{i=1}^{n} |a^*_{ji}|\, |\epsilon_i^{(k)}|$$

$$\leqslant \max_{i=1,...,n} |\epsilon_i^{(k)}| \sum_{i=1}^{n} |a^*_{ji}|$$

$$\leqslant \max_{i=1,...,n} |\epsilon_i^{(k)}| C$$

Thus,

$$\max_{j=1,...,n} |\epsilon_i^{(k+1)}| \leq \max_{i=1,...,n} |\epsilon_i^{(k)}|\, C \leq$$

$$\max_{j=1,...,n} |\epsilon_i^{(k-1)}|\, C^2 \leq \ldots \leq \max_{i=1,...,n} |\epsilon_i^{(0)}|\, C^{k+1}$$

and since $0 < C < 1$, it follows that $\mathbf{x}_k$ converges to the fixed point $\mathbf{x}$ as k approaches $+\infty$, for any initial vector $\mathbf{x}_0$.

2.5.1 Example: Iterative solution of a system of linear equations

We wish to solve the system of linear equations:

$$
\begin{aligned}
8x - 3y + 2z &= 8 \\
4x + 11y - z &= 23 \\
6x + 3y + 12z &= 48
\end{aligned}
$$

Because the diagonal elements are dominant, the Jacobi algorithm is applicable. According to (2.11), we have

$$
\begin{aligned}
x_{k+1} &= \frac{1}{8}\,(3y_k - 2z_k + 8) \\
y_{k+1} &= \frac{1}{11}\,(-4x_k + z_k + 23) \\
z_{k+1} &= \frac{1}{4}\,(-2x_k - y_k + 16)
\end{aligned}
$$

and starting from $x_0 = y_0 = z_0 = 2$, for example, we calculate

$$
\begin{aligned}
x_1 &= 1.25 & \qquad x_2 &= 0.9546 \\
y_1 &= 1.5455 & \qquad y_2 &= 1.8636 \quad \textit{et cetera} \\
z_1 &= 2.5 & \qquad z_2 &= 2.9886
\end{aligned}
$$

The exact solution of the system is given by $x = 1$, $y = 2$, and $z = 3$.

2.5.2 Example: Numerical inverse of a matrix

The Jacobi algorithm can be applied to improve the numerical inverse of a matrix.

Consider a regular matrix **A** and its inverse $\mathbf{A}^{-1}$. Let us assume that we have determined the numerical inverse $\tilde{\mathbf{X}}$, which, due to rounding errors, is only an approximate inverse.

We seek a recurrence formula which allows us to improve $\tilde{\mathbf{X}}$.

We have

$$\tilde{X}A = I + E$$

where **I** is the unit matrix and $\mathbf{E} = [\epsilon_{ij}]$ is the error matrix, with, in general, $|\epsilon_{ij}| \ll 1$. By designating the exact inverse $\mathbf{A}^{-1}$ by **X**, we find

$$\tilde{X} AX = \tilde{X} = X + EX$$

from which we get the equation for **X** in the form of the fixed point:

$$X = -EX + \tilde{X}$$

If

$$\sum_{j=1}^{n} |\epsilon_{ij}| \leqslant C < 1, \qquad i = 1, 2, ..., n,$$

then, the conditions for application of the Jacobi algorithm are satisfied, and we obtain the formula:

$$X_{k+1} = -EX_k + \tilde{X}, \quad k = 0, 1, 2, ...$$

where $\mathbf{X}_0 = \tilde{\mathbf{X}}$.

2.6 EXERCISES

2.6.1 Put the equation $\exp(-x) = \tan x$ into the form of a system of two equations with two unknowns and make one iteration according to the Newton-Raphson method to solve it. Choose initial values by graphing the two curves to obtain the intersection.

2.6.2 Given $f(x,y) = x^3 + y^3 + 3xy$:

- find the extrema of f analytically and determine whether they are maxima, minima, or other types;
- carry out one iteration of the Newton-Raphson method starting from $x_0 = y_0 = -0.9$.

2.6.3 Solve the system of linear equations:

$$\begin{aligned} 7x - 3y + 2z &= 7 \\ 3x + 11y - z &= 22 \\ 5x + 3y + 12z &= 47 \end{aligned}$$

- using the Gaussian elimination method:
- by the iterative method of Jacobi (verify that it can be used) by carrying out two iterations, starting with a suitably chosen initial approximation.

2.6.4 Carry out two iterations to improve the numerical inverse $\tilde{\mathbf{X}}$ of the matrix **A**, where

$$A = \begin{bmatrix} 2 & 2 \\ 2.5 & 3 \end{bmatrix}, \quad \tilde{X} = \begin{bmatrix} 3.1 & -2 \\ -2.4 & 2{,}1 \end{bmatrix}$$

2.6.5 Solve the second-order differential equation:

$$y'' - 3y' + 2y = 1; \quad y(0) = 1, \quad y'(0) = -1$$

according to Picard's method (i.e., reduce the equation to two first-order equations by setting $(y'(x) = z(x))$.

Chapter 3

Difference Equations

3.1 INTRODUCTION

Sampling of analog signals takes on increasing importance in the context of digital signal processing. In this case, the analog signal y (t) is sampled at time intervals ΔT to produce the sequence y_k taken at discrete instants $k\Delta T$, where ΔT is the sampling period and k is an integer.

If the sampled signal (y_k) satisfies a certain recursive relation, for example,

$$y_{k+3} - 5y_{k+2} + 6y_{k+1} + 3y_k = 0, \quad k = 0, 1, 2, \ldots \tag{3.1}$$

then we can progressively calculate all elements y_k, provided that the initial terms y_0, y_1 and y_2 are known. For $y_0 = 1$, $y_1 = -2$, and $y_2 = 0$, for example, according to (3.1), setting $k = 0, 1, 2, \ldots$, we calculate $y_3 = 9$, $y_4 = 51$, $y_5 = 201$ *et cetera.*

A recurrence formula of the form (3.1) is called a third-order ***difference equation***, and a sequence (y_k) satisfying this recurrence is called a ***solution*** of the difference equation.

Because there is a perfect correspondence between an analog function y (t) and the sequence of uniformly sampled values of its ordinate (y_k), the difference equations are amenable to the same methods of solution as those used in the context of differential equations.

3.2 LINEAR DIFFERENCE EQUATIONS WITH CONSTANT COEFFICIENTS

3.2.1 Definitions

Let us consider the difference equations of the form:

$$a_0 y_{k+n} + a_1 y_{k+n-1} + \ldots + a_n y_k = u_k, \quad k = 0, 1, 2, \ldots \tag{3.2}$$

where the constant coefficients $a_0, a_1, a_2, \ldots, a_n$ and the sequence (u_k) are given.

If $u_k = 0$ for each k, equation (3.2) is said to be ***homogeneous***, if not, it is said to be ***nonhomogeneous***.

3.2.2 General solution of the homogeneous equation

Suppose that $(y_k) = (r^k)$ is a soluton of the homogeneous equation associated with (3.2), i.e., of the equation:

$$a_0\, y_{k+n} + a_1\, y_{k+n-1} + \ldots + a_n\, y_n = 0 \tag{3.3}$$

we, therefore, have

$$a_0\, r^{n+k} + a_1\, r^{n+k-1} + \ldots + a_n\, r^k = 0$$

or

$$(a_0\, r^n + a_1\, r^{n-1} + \ldots + a_n)\, r^k = 0$$

It follows that $y_k = r^k$ is a solution of (3.3), if r is a root of the ***characteristic polynomial***

$$P(x) = a_0\, x^n + a_1\, x^{n-1} + \ldots + a_n \tag{3.4}$$

of equation (3.3). Thus, the solutions of (3.3) depend on the roots $r_1, r_2, \ldots, r_n$ of (3.4), and it is in this sense that we distinguish the following three cases:

- ***the roots of the characteristic polynomial*** (3.4) ***are all real and distinct***; each power r_i^k, $i = 1, 2, \ldots, n$, is a solution of (3.3). The general solution of (3.3) is obtained by superposition, i.e.,

  $$y_k = c_1\, r_1^k + c_2\, r_2^k + \ldots + c_n\, r_n^k \tag{3.5}$$

  where c_i, $i = 1, 2, \ldots, n$ are constants;

- ***the characteristic polynomial*** (3.4) ***has complex conjugate roots***; if $\alpha + j\beta$ is a root of (3.4), $\alpha - j\,\beta$ will also be a root. The contribution to the general solution from these two complex conjugate roots is given by

  $$y_k = c_1\,(\alpha + j\beta)^k + c_2\,(\alpha - j\beta)^k \tag{3.6}$$

  In polar coordinates,

  $$\alpha \pm j\beta = \rho \exp(\pm j\varphi) = \rho\,(\cos\varphi \pm j \sin\varphi)$$

  and

  $$(\alpha \pm j\beta)^k = \rho^k (\cos k\,\varphi \pm j \sin k\,\varphi)$$

  from which we find the real part of (3.6):

  $$\begin{aligned} y_k &= c_1'\, \rho^k \cos k\,\varphi + c_2'\, \rho^k \sin k\,\varphi \\ &= c\, \rho^k \cos(k\,\varphi + \psi) \end{aligned} \tag{3.7}$$

- ***the characteristic polynomial*** (3.4) ***has multiple roots***; suppose that $r_2 = r_1$ is a double root. The contribution to the general solution from this double root is then given by

  $$y_k = (c_1 + c_2\, k)\, r_1^k$$

  Likewise, the contribution from the multiple root $r_m = r_{m-1} = r_{m-2} = \ldots = r_1$, $m \leq n$, is given by

  $$y_k = (c_1 + c_2 k + c_3 k^2 + \ldots + c_m\, k^{m-1})\, r_1^k \tag{3.8}$$

3.2.3 General solution of the nonhomogeneous equation

The general solution of the nonhomogeneous equation (3.2) is of the form:

$$(y_k) = (h_k) + (p_k) \tag{3.9}$$

where (h_k) is the general solution of the ***homogeneous*** equation (3.3) associated with (3.2), and (p_k) is a ***particular*** solution of the nonhomogeneous equation (3.2).

3.2.4 Graphical representation of a linear difference equation with constant coefficients

The graphical representation of a linear difference equation is based on three elements:

- ***the additive element***, with the output signal equal to the sum of the input signals (see fig. 3.1, for example);

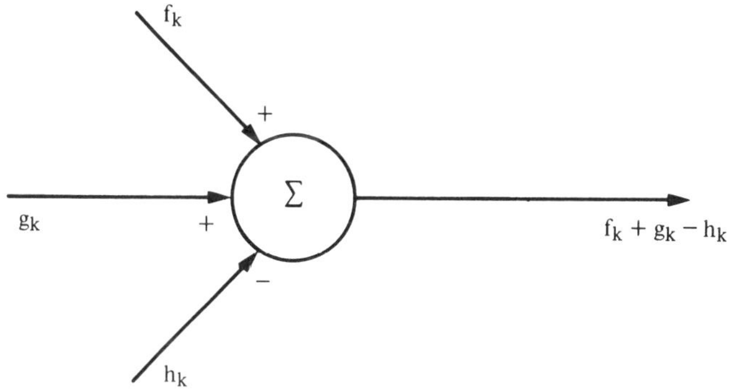

Fig. 3.1

- ***the multiplicative element***, with an output signal which is the input signal multiplied by the coefficient indicated (see fig. 3.2, for example);

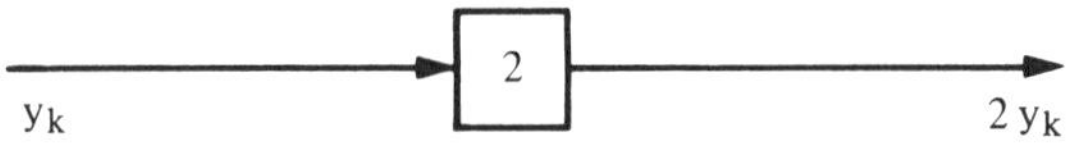

Fig. 3.2

- ***the delay element***, with an output signal which is the input signal delayed by one period (see fig. 3.3, for example).

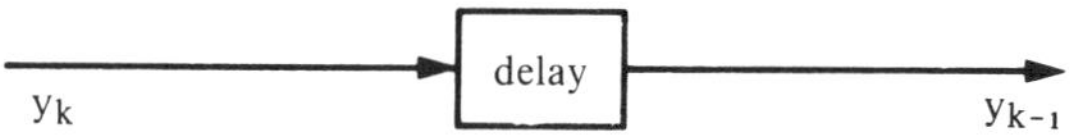

Fig. 3.3

3.2.5 Example: Schematic representation

The difference equation:

$$y_{k+3} - 2y_{k+2} - y_{k+1} + 2y_k = 0$$

can be written in the form:

$$y_{k+3} = 2y_{k+2} + y_{k+1} - 2y_k$$

from which we get the schematic representation of figure 3.4.

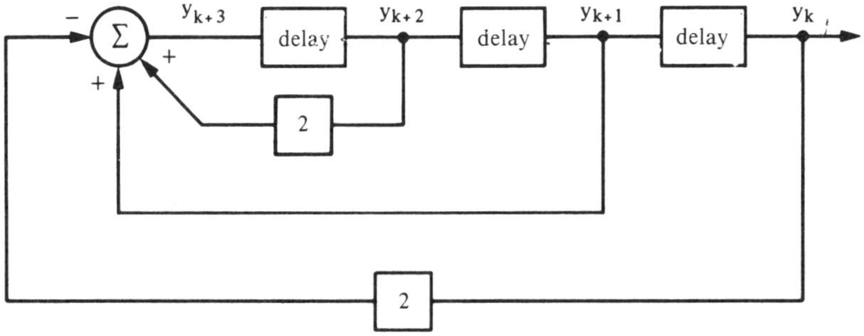

Fig. 3.4

We represent systems of differential equations in an analogous fashion (see exercise 3.4.2).

3.2.6 Example: Electrical network

Let us consider the electrical network of figure 3.5, in which R_1 and R_2 are constant resistances, V is the voltage supplied by a generator, and I_k is the current in the *k*th loop.

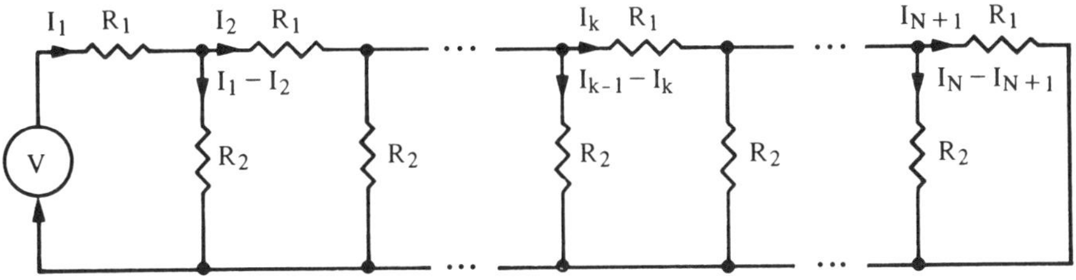

Fig. 3.5

According to Kirchhoff's laws, we find for the first loop:

$$I_1 R_1 + (I_1 - I_2) R_2 - V = 0 \qquad \text{(a)}$$

for the *k*th loop, $2 \leq k \leq N$, we have

$$I_k R_1 + (I_k - I_{k+1}) R_2 - (I_{k-1} - I_k) R_2 = 0 \qquad \text{(b)}$$

and for the (N+1)th loop, we have

$$I_{N+1} R_1 - (I_N - I_{N+1}) R_2 = 0 \qquad \text{(c)}$$

Writing equation (b) in the form:

$$I_{k+1} - \left(2 + \frac{R_1}{R_2}\right) I_k + I_{k-1} = 0 \qquad \text{(d)}$$

we obtain a linear homogeneous difference equation with constant coefficients. Setting $I_k = r^k$, we find its characteristic polynomial:

$$P(r) = r^2 - \left(2 + \frac{R_1}{R_2}\right) r + 1$$

the roots of which are

$$r_{1,2} = 1 + \frac{R_1}{2\,R_2} \pm \sqrt{\frac{R_1}{R_2} + \frac{R_1^2}{4\,R_2^2}} \tag{e}$$

According to equation (3.5), the general solution of (d) is of the form:

$$I_k = c_1\, r_1^k + c_2\, r_2^k, \quad k = 1, 2, \ldots, N + 1 \tag{f}$$

To determine the constants c_1 and c_2, we proceed in the following manner. According to equation (a) we have

$$I_2 = \frac{I_1\,(R_1 + R_2) - V}{R_2} \tag{g}$$

and from equation (c), we find

$$I_N = \frac{I_{N+1}\,(R_1 + R_2)}{R_2} \tag{h}$$

Substituting the current from expression (f) in equations (g) and (h), we obtain, respectively,

$$c_1\, r_1^2 + c_2\, r_2^2 = \frac{(c_1\, r_1 + c_2\, r_2)\,(R_1 + R_2) - V}{R_2}$$

and

$$c_1\, r_1^N + c_2\, r_2^N = \frac{(c_1\, r_1^{N+1} + c_2\, r_2^{N+1})\,(R_1 + R_2)}{R_2}$$

from which we get c_1 and c_2. After substitution of c_1 and c_2 in formula (f), we find the current I_k in the *k*th loop as a function of the resistances R_1 and R_2, and of the voltage V.

3.3 NUMERICAL SOLUTION OF THE ALGEBRAIC EQUATION

Let us consider the ***algebraic equation***

$$P(x) = a_0\, x^n + a_1\, x^{n-1} + \ldots + a_n = 0 \tag{3.10}$$

in which the coefficients a_i are real constants, $i = 0, 1, \ldots, n$, and $a_0 \neq 0$. In view of determining its roots $x_1, x_2, \ldots, x_n$, we distinguish between the two following cases:

- ***the roots of equation*** (3.10) are all ***real and distinct***; considering the associated difference equation:

$$a_0 y_{k+n} + a_1 y_{k+n-1} + \ldots + a_n y_k = 0, \quad k = 0, 1, 2, \ldots \tag{3.11}$$

constructed with the aid of the coefficients of (3.10), we find that equation (3.10) is the characteristic equation of (3.11). The general solution of (3.11) is given by relation (3.5):

$$y_k = c_1 x_1^k + c_2 x_2^k + \ldots + c_n x_n^k \tag{3.12}$$

where the constants $c_1, c_2, \ldots, c_n$ are determined by the initial conditions:

$$\begin{aligned} y_0 &= c_1 + c_2 + \ldots + c_n \\ y_1 &= c_1 x_1 + c_2 x_2 + \ldots + c_n x_n \\ &\vdots \\ y_{n-1} &= c_1 x_1^{n-1} + c_2 x_2^{n-1} + \ldots + c_n x_n^{n-1} \end{aligned}$$

Suppose that x_1 is the root of largest absolute value of (3.10) ($|x_1| > |x_i|$, $i = 2, 3, \ldots, n$). Under the condition that $c_1 \neq 0$, we shall demonstrate that the ratio of two consecutive terms y_{k+1} and y_k of the solution (3.12) converges to x_1 as k approaches ∞:

$$x_1 = \lim_{k \to \infty} \frac{y_{k+1}}{y_k} \tag{3.13}$$

The algorithm determined by formula (3.13) is called ***Bernoulli's real method.***

To verify formula (3.13), let us write (3.12) in the following way:

$$y_k = x_1^k \left\{ c_1 + c_2 \left(\frac{x_2}{x_1} \right)^k + \ldots + c_n \left(\frac{x_n}{x_1} \right)^k \right\}$$

and, thus,

$$y_{k+1} = x_1^{k+1} \left\{ c_1 + c_2 \left(\frac{x_2}{x_1} \right)^{k+1} + \ldots + c_n \left(\frac{x_n}{x_1} \right)^{k+1} \right\}$$

from which we get

$$\frac{y_{k+1}}{y_k} = x_1 \frac{c_1 + c_2 \left(\frac{x_2}{x_1} \right)^{k+1} + \ldots + c_n \left(\frac{x_n}{x_1} \right)^{k+1}}{c_1 + c_2 \left(\frac{x_2}{x_1} \right)^{k} + \ldots + c_n \left(\frac{x_n}{x_1} \right)^{k}}$$

Since $|x_1| > |x_i|$, $i = 2, 3, \ldots n$, it is easily seen that y_{k+1}/y_k converges to x_1 as k approaches ∞. If $c_1 = 0$, which can occur for arbitrary choice of initial conditions, but $c_2 \neq 0$, the limit (3.13) is equal to the second largest root x_2 of equation (3.10) (with the tacit supposition that $|x_2| > |x_i|, i = 3, 4, \ldots, n$).

To calculate the remaining roots of equation (3.10), we first divide the polynomial P (x) by the monome $x - x_1$. Thus, we obtain a polynomial with roots $x_2, x_3, \ldots, x_n$; and because all these roots are distinct, we again apply Bernoulli's real algorithm to determine x_2 (we assume that $|x_2| > |x_3| > \ldots > |x_n|$). We find the remaining roots in an analogous manner.

In the case in which x_1 is a multiple root of equation (3.10), formula (3.13) is also applicable. After determining x_1, we continue in the manner that has just been described.

- ***equation*** (3.10) ***has complex conjugate roots***; let us assume that the two roots of largest absolute value of equation (3.10) are complex conjugate: $z_1 = \rho \exp(j\varphi)$ and $z_2 = \rho \exp(-j\varphi)$: the other roots can be real or complex conjugate. Equation (3.10) is again the characteristic equation of the difference equation (3.11) whose general solution is of the form (see (3.5) and (3.7))

$$y_k = c\,\rho^k \cos(k\varphi + \psi) + \ldots \tag{3.14}$$

In this case, the sequence formed of the quotients (y_{k+1}/y_k) which we have just considered in the real method (3.13), does not converge, but will oscillate around a certain value. On the other hand, we have

$$\rho = \lim_{k \to \infty} \sqrt{\frac{D_{k+1}}{D_k}} \tag{3.15}$$

and

$$\cos\varphi = \lim_{k \to \infty} \frac{E_k}{2\sqrt{D_{k+1} D_k}} \tag{3.16}$$

with

$$D_k = \begin{vmatrix} y_{k-1} & y_{k-2} \\ y_k & y_{k-1} \end{vmatrix}$$

$$E_k = \begin{vmatrix} y_k & y_{k-2} \\ y_{k+1} & y_{k-1} \end{vmatrix}$$

where the y_k are solutions (3.14) of equation (3.11). The algorithm determined by formulas (3.15) and (3.16) is called the ***complex Bernoulli method***.

Let us demonstrate formulas (3.15) and (3.16). According to the assumption that $\rho > | x_i |$, i = 3, 4, ..., n, we can approximate y_k (see (3.14) for large k by $c\,\rho^k \cos(k\varphi + \psi)$, which entails the following approximations:

$$D_k \approx c^2 \rho^{2(k-1)} \{\cos^2[(k-1)\varphi + \psi] - \cos(k\varphi + \psi)\cos[(k-2)\varphi + \psi]\}$$
$$= c^2 \rho^{2(k-1)} \{1/2\,(1 + \cos[2(k-1)\varphi + 2\psi])$$
$$-1/2\,(\cos 2\varphi + \cos[2(k-1)\varphi + 2\psi])\}$$
$$= c^2 \rho^{2(k-1)}\, 1/2\,(1 - \cos 2\varphi) = c^2 \rho^{2(k-1)} \sin^2\varphi$$

from which we get $D_{k+1}/D_k \approx \rho^2$, and thus equation (3.15).
Likewise, we have

$$E_k \approx c^2 \rho^{2k-1} \{\cos[k\varphi + \psi]\cos[(k-1)\varphi + \psi]$$
$$- \cos[(k+1)\varphi + \psi]\cos[(k-2)\varphi + \psi]\}$$
$$= c^2 \rho^{2k-1}\, 1/2\, \{\cos[(2k-1)\varphi + 2\psi] + \cos\varphi$$
$$- \cos[(2k-1)\varphi + 2\psi] - \cos 3\varphi\}$$
$$= c^2 \rho^{2k-1}\, 1/2\, \{\cos\varphi - \cos 3\varphi\} = c^2 \rho^{2k-1}\, 2\cos\varphi \sin^2\varphi$$

from which we get

$$\frac{E_k}{2\sqrt{D_{k+1} D_k}} \approx \frac{c^2 \rho^{2k-1}\, 2\cos\varphi\sin^2\varphi}{2\sqrt{c^2 \rho^{2(k-1)} \sin^2\varphi\, c^2 \rho^{2k} \sin^2\varphi}}$$

and thus equation (3.16).

3.3.1 Example: Simple roots

Given the algebraic equation:

$$x^3 - 6x^2 + 11x - 6 = 0 \tag{a}$$

for which we wish to calculate the root of largest absolute value. Because its roots are $x_1 = 3$, $x_2 = 2$ and $x_3 = 1$, we can do this by the real Bernoulli method (3.13).

We associate equation (a) with the difference equation:

$$y_{k+3} - 6y_{k+2} + 11y_{k+1} - 6y_k = 0$$

from which we get

$$y_{k+3} = 6y_{k+2} - 11y_{k+1} + 6y_k$$

Starting with the initial conditions $y_0 = 0$, $y_1 = 0$, $y_2 = 1$, we calculate the values of table 3.6.

Table 3.6

	k	y_{0+k}	y_{1+k}	y_{2+k}	y_{3+k}	y_{k+1}/y_k
Weighting coefficients →		6	− 11	6		
Initial conditions →	0	0	0	1	6	4,16
	1	0	1	6	25	3,6
	2	1	6	25	90	3,34
	3	6	25	90	301	3,20
	4	25	90	301	966	3,13
	5	90	301	966	3 025	3,08
	6	301	966	3 025	9 330	3,05
	7	966	3 025	9 330	28 501	

Thus, we find that y_{k+1}/y_k converges to $x_1 = 3$ as k approaches ∞.

3.3.2 Example: Complex conjugate roots

Given the equation:

$$x^3 - 5x^2 + 12x - 8 = 0 \qquad \text{(a)}$$

Its roots are $z_1 = 2 + j2$, $z_2 = 2 - j2$ and $x_3 = 1$. To calculate the complex conjugate roots of largest absolute value, we must apply the complex Bernoulli method (3.14) and (3.15).

We associate the difference equation with equation (a):

$$y_{k+3} - 5y_{k+2} + 12y_{k+1} - 8y_k = 0$$

from which we get

$$y_{k+3} = 5y_{k+2} - 12y_{k+1} + 8y_k$$

Starting with the initial conditions $y_0 = 0$, $y_1 = 0$, $y_2 = 1$, we calculate the values of table 3.7.

We have, for example,

$$D_{12} = \begin{vmatrix} 19\,661 & 3\,277 \\ 52\,429 & 19\,661 \end{vmatrix} = 214\,745\,088$$

$$D_{11} = \begin{vmatrix} 3\,277 & -819 \\ 19\,661 & 3\,277 \end{vmatrix} = 26\,841\,088$$

$$E_{11} = \begin{vmatrix} 19\,661 & -819 \\ 52\,429 & 3\,277 \end{vmatrix} = 107\,368\,448$$

thus,

$$\sqrt{\frac{D_{12}}{D_{11}}} = 2.8286 \quad \text{and} \quad \frac{E_{11}}{2\sqrt{D_{12}\,D_{11}}} = 0.7071$$

we find

$$\rho = \lim_{k\to\infty} \sqrt{\frac{D_{k+1}}{D_k}} = 2.8284$$

and

$$\cos\varphi = \lim_{k\to\infty} \frac{E_k}{2\sqrt{D_{k+1}\,D_k}} = 0.7071$$

i.e., $z_1 = 2 + j2$ and $z_2 = 2 - j2$.

Table 3.7

	k	y_{0+k}	y_{1+k}	y_{2+k}	y_{3+k}
Weighting coefficients →		8	−12	5	
Initial conditions →	0	0	0	1	5
	1	0	1	5	13
	2	1	5	13	13
	3	5	13	13	−51
	4	13	13	−51	−307
	5	13	−51	−307	−819
	6	−51	−307	−819	−819
	7	−307	−819	−819	3 277
	8	−819	−819	3 277	19 661
	9	−819	3 277	19 661	52 429

3.4 EXERCISES

3.4.1 Given the linear difference equation:

$$x_{k+3} - 3\,x_{k+2} + 3\,x_{k+1} - x_k = 0$$

- draw a schematic of the equation;
- find the general solution and deduce an explicit formula for the sum of the first k integers:

$$S_k = \sum_{n=0}^{k} n$$

3.4.2 Draw the schematic of a system of the two difference equations:

$$x_{k+1} = x_k + y_k$$

$$y_{k+1} = x_k - y_k$$

and determine the solution in explicit form, satisfying the initial conditions $x_0 = 1$, $y_0 = 1$.

3.4.3 Consider the diagram of figure 3.8.

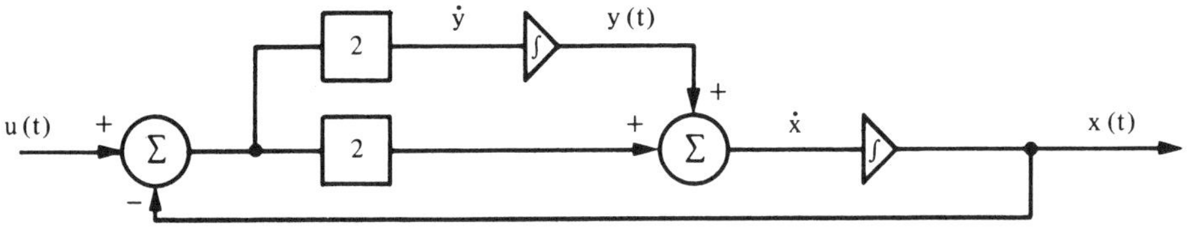

Fig. 3.8

- Find the differential equations corresponding to the diagram.
- The functions u (t), x (t), and y (t) are sampled uniformly with the period T, and the series (u_k), (x_k), and (y_k) are thus obtained. Approximate the derivative $\dot{x}(t)$ by $(x_{k+1} - x_k)/T$ and the derivative $\dot{y}(t)$ by $(y_{k+1} - y_k)\,T$, and establish the corresponding difference equations.
- Draw the schematic corresponding to the difference equations, and compare the analog and sampled data circuits.

3.4.4 Calculate the complex conjugate roots z_1 and z_2 of largest absolute value of the polynomial:

$$P(x) = x^4 - 8\,x^3 + 26\,x^2 - 8\,x + 25$$

using Bernoulli's method. Then,

- choose the initial values $y_k = 1$, $k = 0, 1, 2, 3$, and verify that the sequence (y_{k+1}/y_k) of the real method does not converge;
- with the same initial values, apply the complex method;
- calculate the other roots of P (x) algebraically.

3.4.5 Knowing that the root of smallest absolute value is real, find this root using Bernoulli's method, then determine the remaining roots of

$$P(x) = x^3 - 3.5\,x^2 + 6\,x + 4$$

Chapter 4

Eigenvalues and Eigenvectors

4.1 INTRODUCTION

For many applications of matrix calculations, the eigenvalues and eigenvectors play an important role, and it is of great practical interest to know the procedures that allow us to determine or to make an approximation of their numerical values.

We shall illustrate this type of problem by studying the motion of a transmission shaft that is fixed at one end and free at the other, and represented by three equidistant disks with equal moments of inertia J. Without taking friction into account, we assume that the disks are connected to each other by spiral springs of zero mass and torsion characteristic C.

If φ_1, φ_2, and φ_3 designate the respective angular movements of the three disks from the equilibrium position (see fig. 4.1), we obtain

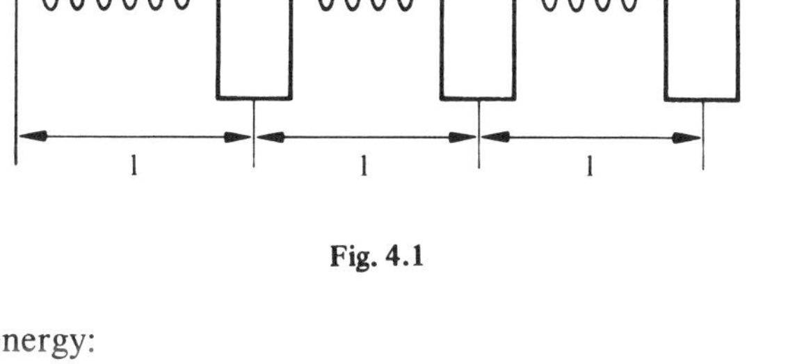

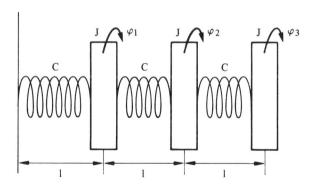

Fig. 4.1

for the kinetic energy:

$$T = \frac{1}{2} J (\dot{\varphi}_1^2 + \dot{\varphi}_2^2 + \dot{\varphi}_3^2)$$

and for the potential energy:

$$U = \frac{C}{2} (\varphi_1^2 + (\varphi_2 - \varphi_1)^2 + (\varphi_3 - \varphi_2)^2)$$

The Lagrange equations:

$$\frac{\partial L}{\partial \varphi_k} - \frac{d}{dt} \frac{\partial L}{\partial \dot{\varphi}_k} = 0, \qquad k = 1, 2, 3$$

where L=T–U provide the three differential equations of motion:

$$2\,\varphi_1 - \varphi_2 = -\frac{J}{C}\ddot{\varphi}_1$$

$$-\varphi_1 + 2\varphi_2 - \varphi_3 = -\frac{J}{C}\ddot{\varphi}_2$$

$$-\varphi_2 + \varphi_3 = -\frac{J}{C}\ddot{\varphi}_3$$

or, in matrix form:

$$\mathbf{S}\,\varphi = -\frac{J}{C}\ddot{\varphi} \tag{4.1}$$

with

$$\mathbf{S} = \begin{bmatrix} 2 & -1 & 0 \\ -1 & 2 & -1 \\ 0 & -1 & 1 \end{bmatrix}, \qquad \varphi = \begin{bmatrix} \varphi_1 \\ \varphi_2 \\ \varphi_3 \end{bmatrix}, \qquad \ddot{\varphi} = \begin{bmatrix} \ddot{\varphi}_1 \\ \ddot{\varphi}_2 \\ \ddot{\varphi}_3 \end{bmatrix}$$

By setting $\varphi = \varphi_0 \exp(j\omega t)$, which corresponds to the hypothesis of a sinusoidal oscillation, system (4.1) becomes, after division by $\exp(j\omega t)$:

$$\mathbf{S}\,\varphi_0 = \lambda\,\varphi_0 \tag{4.2}$$

where

$$\lambda = \frac{J\,\omega^2}{C}$$

Because we are interested in the values of λ for which the system (4.2) has nonzero solutions φ_0, we are confronted with the standard problem, which is to determine the eigenvalues and the eigenvectors of a symmetric matrix.

4.2 GENERALITIES

Assume that $\mathbf{A}$ is a square matrix with n lines and n columns whose coefficients are real numbers.

A scalar λ is called an ***eigenvalue*** of $\mathbf{A}$, if there exists a nonzero column vector $\mathbf{x}$ such that $\mathbf{A x} = \lambda \mathbf{x}$. Any vector satisfying this condition is said to be an ***eigenvector*** of $\mathbf{A}$ associated with the eigenvalue λ. The set of all the eigenvalues of $\mathbf{A}$ is called the ***spectrum*** of $\mathbf{A}$.

The matrix $\mathbf{A} - \lambda\, \mathbf{I}$ is called the ***characteristic matrix*** of $\mathbf{A}$. The polynomial $\Phi(\lambda) = \det(\mathbf{A} - \lambda\, \mathbf{I})$ is called the ***characteristic polynomial*** of $\mathbf{A}$, and the equation $\Phi(\lambda) = 0$ is called the ***characteristic equation*** of $\mathbf{A}$.

According to the theory of systems of linear equations, we know that a scalar λ is an eigenvalue of $\mathbf{A}$ if and only if λ is a root of the characteristic polynomial $\Phi(\lambda)$ of $\mathbf{A}$.

The determinant:

$$\Phi(\lambda) = \det[\mathbf{A} - \lambda \mathbf{I}] = \begin{vmatrix} a_{11} - \lambda & a_{12} & \dots & a_{1n} \\ a_{21} & a_{22} - \lambda & & a_{2n} \\ \vdots & & & \\ a_{n1} & & & a_{nn} - \lambda \end{vmatrix}$$

is a polynomial of degree n, and each polynomial of degree n has at least one root and no more than n distinct roots. Thus, $\mathbf{A}$ has at least one and no more than n distinct eigenvalues. If no assumptions are made about $\mathbf{A}$, the eigenvalues can be complex numbers.

Without demonstration, we mention the following facts:

- the eigenvectors $\mathbf{x}_1, \mathbf{x}_2, \dots, \mathbf{x}_r$ of a matrix $\mathbf{A}$ associated with the distinct eigenvalues $\lambda_1, \lambda_2, \dots, \lambda_r$, are linearly independent;
- if $\mathbf{S}$ is a ***symmetric*** matrix, its eigenvalues are real, and its eigenvectors corresponding to distinct eigenvalues are orthogonal.

Given a symmetric matrix $\mathbf{S}$ with n distinct eigenvalues $\lambda_1, \lambda_2, \dots, \lambda_n$; if the associated eigenvectors $\mathbf{x}_1, \mathbf{x}_2, \dots, \mathbf{x}_n$ are ***normalized***, an orthonormal base is obtained. A square matrix $\mathbf{P}$ is called ***orthogonal***, if $\mathbf{P}^T\mathbf{P} = \mathbf{P}\mathbf{P}^T = \mathbf{I}$. By constructing the orthogonal matrix $\mathbf{P} = [\mathbf{x}_1, \mathbf{x}_2, \dots, \mathbf{x}_n]$ whose columns are the n eigenvectors, we have

$$\begin{aligned}\mathbf{SP} &= [\mathbf{Sx}_1, \mathbf{Sx}_2, \ldots, \mathbf{Sx}_n] \\ &= [\lambda_1 \mathbf{x}_1, \lambda_2 \mathbf{x}_2, \ldots, \lambda_n \mathbf{x}_n] \\ &= [\mathbf{x}_1, \mathbf{x}_2, \ldots, \mathbf{x}_n] \begin{bmatrix} \lambda_1 & 0 & \ldots 0 \\ 0 & \lambda_2 & \\ \vdots & & \\ 0 & & \lambda_n \end{bmatrix} = \mathbf{P}\,\Lambda,\end{aligned}$$

where

$$\Lambda = \begin{bmatrix} \lambda_1 & 0 & \ldots 0 \\ 0 & \lambda_2 & \\ \vdots & & \\ 0 & & \lambda_n \end{bmatrix}$$

Thus,

$$\mathbf{S} = \mathbf{P}\,\Lambda\,\mathbf{P}^{-1} = \mathbf{P}\,\Lambda\,\mathbf{P}^{T} \tag{4.3}$$

or

$$\Lambda = \mathbf{P}^{-1}\,\mathbf{S}\,\mathbf{P} = \mathbf{P}^{T}\,\mathbf{S}\,\mathbf{P} \tag{4.4}$$

Equation (4.3) or (4.4) represents the ***diagonalization*** of **S**.

4.3 ITERATED POWER ALGORITHM

Let **S** be a symmetric matrix with distinct eigenvalues $\lambda_1, \lambda_2, \ldots, \lambda_n$. Because their eigenvectors $\mathbf{x}_1, \mathbf{x}_2, \ldots, \mathbf{x}_n$ are orthogonal, each vector **v** with n components can be represented as a linear combination of these eigenvectors:

$$\mathbf{v} = \sum_{i=1}^{n} \alpha_i\, \mathbf{x}_i$$

where the coefficients α_i, i = 1, 2, . . ., n, are real numbers. Because the eigenvectors are orthogonal, we have

$$\alpha_k = \frac{\mathbf{v}^T\, \mathbf{x}_k}{\|\mathbf{x}_k\|^2}, \qquad k = 1, 2, \ldots, n$$

and, further, if the eigenvectors $\mathbf{x}_k$ are normalized, we obtain simply

$$\alpha_k = \mathbf{v}^T \mathbf{x}_k, \qquad k = 1, 2, \ldots, n$$

Let us now consider the series of iterated powers:

$$\mathbf{Sv}, \mathbf{S}^2\mathbf{v}, \ldots, \mathbf{S}^k\mathbf{v}, \ldots:$$

$$\mathbf{Sv} = \sum_{i=1}^{n} \alpha_i \mathbf{Sx}_i = \sum_{i=1}^{n} \alpha_i \lambda_i \mathbf{x}_i$$

$$\mathbf{S}^2\mathbf{v} = \sum_{i=1}^{n} \alpha_i \lambda_i \mathbf{Sx}_i = \sum_{i=1}^{n} \alpha_i \lambda_i^2 \mathbf{x}_i$$

$$\vdots$$

$$\mathbf{S}^k\mathbf{v} = \sum_{i=1}^{n} \alpha_i \lambda_i^{k-1} \mathbf{Sx}_i = \sum_{i=1}^{n} \alpha_i \lambda_i^k \mathbf{x}_i$$

$$\vdots$$

By writing

$$\mathbf{S}^k\mathbf{v} = \lambda_1^k \left\{ \alpha_1 \mathbf{x}_1 + \alpha_2 \left(\frac{\lambda_2}{\lambda_1} \right)^k \mathbf{x}_2 + \ldots + \alpha_n \left(\frac{\lambda_n}{\lambda_1} \right)^k \mathbf{x}_n \right\} \tag{4.5}$$

and assuming that $|\lambda_1| > |\lambda_2| > \ldots > |\lambda_n|$, we obtain for the eigenvalue of largest absolute value if $\alpha_1 \neq 0$:

$$|\lambda_1| = \lim_{k \to \infty} \frac{\|\mathbf{S}^k\mathbf{v}\|}{\|\mathbf{S}^{k-1}\mathbf{v}\|} \tag{4.6}$$

To determine the eigenvector associated with λ_1, we assume first that $\lambda_1 > 0$. In this case, equation (4.5) implies, if $\alpha_1 \neq 0$:

$$\lim_{k \to \infty} \frac{\mathbf{S}^k\mathbf{v}}{\|\mathbf{S}^k\mathbf{v}\|} = \frac{\alpha_1 \mathbf{x}_1}{\|\alpha_1 \mathbf{x}_1\|} \tag{4.7a}$$

i.e., we find the normalized eigenvector that we sought. If $\lambda_1 < 0$, it follows analogously that

$$\lim_{k \to \infty} \frac{(-1)^k \mathbf{S}^k\mathbf{v}}{\|\mathbf{S}^k\mathbf{v}\|} = \frac{\alpha_1 \mathbf{x}_1}{\|\alpha_1 \mathbf{x}_1\|} \tag{4.7b}$$

Therefore, λ_1 is positive, if the corresponding components of $\mathbf{S}^k\mathbf{v}$ and $\mathbf{S}^{k-1}\mathbf{v}$ have the same sign, otherwise λ_1 is negative.

The procedure that we have just developed is called the ***iterated power algorithm***. By normalizing the vectors $\mathbf{S}^k\,\mathbf{v}$ at each stage, we avoid their components becoming too large.

After determining λ_1 and $\mathbf{x}_1$, we shall determine λ_2 and its associated normalized eigenvector by again applying the iterated power algorithm to a matrix $\mathbf{S}^*$, which has the eigenvalues $0, \lambda_2, \lambda_3, \ldots, \lambda_n$. $\mathbf{S}^*$ is defined as follows:

$$\mathbf{S}^* = \mathbf{S} - \lambda_1\,\mathbf{x}_1\,\mathbf{x}_1^T \tag{4.8}$$

where $\mathbf{x}_1$ is the normalized eigenvector of $\mathbf{S}$ associated with λ_1.

We verify that $\mathbf{S}^*$ has the eigenvalues $0, \lambda_2, \lambda_3, \ldots, \lambda_n$. Because the multiplication of matrices is associative and $\mathbf{x}_1^T\mathbf{x}_1 = 1$, we find

$$\begin{aligned}\mathbf{S}^*\mathbf{x}_1 &= \mathbf{S}\mathbf{x}_1 - \lambda_1\,(\mathbf{x}_1\,\mathbf{x}_1^T)\,\mathbf{x}_1 = \mathbf{S}\mathbf{x}_1 - \lambda_1\,\mathbf{x}_1\,(\mathbf{x}_1^T\,\mathbf{x}_1)\\ &= \lambda_1\,\mathbf{x}_1 - \lambda_1\,\mathbf{x}_1 = \mathbf{0}\end{aligned}$$

Thus, 0 is an eigenvalue of $\mathbf{S}^*$. For $k \neq 1$, we have $\mathbf{x}_1^T\mathbf{x}_k = 0$, from which we get

$$\begin{aligned}\mathbf{S}^*\mathbf{x}_k &= \mathbf{S}\mathbf{x}_k - \lambda_1\,(\mathbf{x}_1\,\mathbf{x}_1^T)\,\mathbf{x}_k\\ &= \lambda_k\,\mathbf{x}_k - \lambda_1\,\mathbf{x}_1\,(\mathbf{x}_1^T\,\mathbf{x}_k) = \lambda_k\,\mathbf{x}_k\end{aligned}$$

which shows that the λ_k, $k \neq 1$, are eigenvalues of $\mathbf{S}^*$.

By applying the iterated power algorithm to $\mathbf{S}^*$, we can determine λ_2 and $\mathbf{x}_2$. We proceed analogously to calculate the remaining eigenvalues and their associated eigenvectors.

4.3.1 Example: Iterative calculation of eigenvalues and eigenvectors

Let us consider the symmetric matrix:

$$\mathbf{S} = \begin{bmatrix} 4 & 2 & 2 \\ 2 & 5 & 1 \\ 2 & 1 & 5 \end{bmatrix}$$

Its eigenvalues are

$$\lambda_1 = 8, \qquad \lambda_2 = 4, \qquad \lambda_3 = 2$$

and their associated normalized eigenvectors are

$$\mathbf{x}_1 = \frac{1}{\sqrt{3}}\begin{bmatrix}1\\1\\1\end{bmatrix}, \quad \mathbf{x}_2 = \frac{1}{\sqrt{2}}\begin{bmatrix}0\\-1\\1\end{bmatrix}, \quad \mathbf{x}_3 = \frac{1}{\sqrt{6}}\begin{bmatrix}-2\\1\\1\end{bmatrix}$$

Starting with

$$\mathbf{v} = \begin{bmatrix}1\\2\\1\end{bmatrix}$$

for example, we calculate according to the iterated power algorithm applied to **S**:

$$\mathbf{S}\mathbf{v} = \begin{bmatrix}10\\13\\9\end{bmatrix}, \quad \mathbf{S}^2\mathbf{v} = \begin{bmatrix}84\\94\\78\end{bmatrix}, \quad \mathbf{S}^3\mathbf{v} = \begin{bmatrix}680\\716\\652\end{bmatrix}, \quad \mathbf{S}^4\mathbf{v} = \begin{bmatrix}5456\\5592\\5336\end{bmatrix}, \textit{et cetera}$$

Following formula (4.6), we obtain for the largest absolute value of the eigenvalues:

$$|\lambda_1| \approx \frac{\|\mathbf{S}^4\mathbf{v}\|}{\|\mathbf{S}^3\mathbf{v}\|} = \frac{(5456^2 + 5592^2 + 5336^2)^{1/2}}{(680^2 + 716^2 + 652^2)^{1/2}} = \frac{9461.04}{1183.28} = 8.00$$

and since the corresponding components of the iterated vectors have the same sign, we have $\lambda_1 = 8$. For an approximation of its associated normalized eigenvector, we find, according to formula (4.7a):

$$\mathbf{x}_1 \approx \frac{\mathbf{S}^4\mathbf{v}}{\|\mathbf{S}^4\mathbf{v}\|} = \frac{1}{(5456^2 + 5592^2 + 5336^2)^{1/2}}\begin{bmatrix}5456\\5592\\5336\end{bmatrix} = \begin{bmatrix}0.58\\0.59\\0.56\end{bmatrix}$$

To calculate λ_2 and $\mathbf{x}_2$, we first transform the matrix **S** into **S***, where **S*** is defined by equation (4.8). This gives

$$\mathbf{S}^* = \mathbf{S} - \lambda_1 \mathbf{x}_1 \mathbf{x}_1^{\mathrm{T}} = \begin{bmatrix} 4 & 2 & 2 \\ 2 & 5 & 1 \\ 2 & 1 & 5 \end{bmatrix} - \frac{8}{3} \begin{bmatrix} 1 \\ 1 \\ 1 \end{bmatrix} [1 \quad 1 \quad 1] =$$

$$= \begin{bmatrix} 4 & 2 & 2 \\ 2 & 5 & 1 \\ 2 & 1 & 5 \end{bmatrix} - \frac{1}{3} \begin{bmatrix} 8 & 8 & 8 \\ 8 & 8 & 8 \\ 8 & 8 & 8 \end{bmatrix} = \begin{bmatrix} 4/3 & -2/3 & -2/3 \\ -2/3 & 7/3 & -5/3 \\ -2/3 & -5/3 & 7/3 \end{bmatrix}$$

hence, we again apply the iterated power algorithm to $\mathbf{S}^*$, starting, for example, from

$$\mathbf{w} = \begin{bmatrix} 0 \\ 0 \\ 1 \end{bmatrix}$$

We calculate

$$\mathbf{S}^*\mathbf{w} = \frac{1}{3} \begin{bmatrix} -2 \\ -5 \\ 7 \end{bmatrix}, \quad \mathbf{S}^{*2}\mathbf{w} = \frac{1}{3^2} \begin{bmatrix} -12 \\ -66 \\ 78 \end{bmatrix}, \quad \mathbf{S}^{*3}\mathbf{w} = \frac{1}{3^3} \begin{bmatrix} -72 \\ -828 \\ 900 \end{bmatrix}$$

$$\mathbf{S}^{*4}\mathbf{w} = \frac{1}{3^4} \begin{bmatrix} -432 \\ -10152 \\ 10584 \end{bmatrix} \textit{ et cetera}$$

thus,

$$|\lambda_2| \approx \frac{\|\mathbf{S}^{*4}\mathbf{w}\|}{\|\mathbf{S}^{*3}\mathbf{w}\|} = \frac{181.14}{45.37} = 3.9925$$

and taking into account that corresponding components of the iterated vectors do not change signs, we have $\lambda_2 \approx 4$.

For $\mathbf{x}_2$, we find

$$\mathbf{x}_2 \approx \frac{\mathbf{S}^{*4}\mathbf{w}}{\|\mathbf{S}^{*4}\mathbf{w}\|} = \begin{bmatrix} -0.03 \\ -0.69 \\ 0.72 \end{bmatrix}$$

The calculation of λ_3 and $\mathbf{x}_3$ is the object of exercise 4.4.4.

4.4 EXERCISES

4.4.1 Determine the eigenvalues and the eigenvectors of the matrix below (see section 4.1):

$$\mathbf{S} = \begin{bmatrix} 2 & -1 & 0 \\ -1 & 2 & -1 \\ 0 & -1 & 1 \end{bmatrix}$$

4.4.2 Give the differential equations of motion of the two carts represented in figure 4.2.

- Write the equations in matrix form $\ddot{\mathbf{x}} = \mathbf{A}\mathbf{x}$, where

$$\mathbf{x} = \begin{bmatrix} x \\ y \end{bmatrix}$$

- Set $\mathbf{x} = \mathbf{x}_0 \exp(j\omega t)$ and determine the resonant frequencies and the two vectors $\mathbf{x}_0$ of the associated amplitudes.

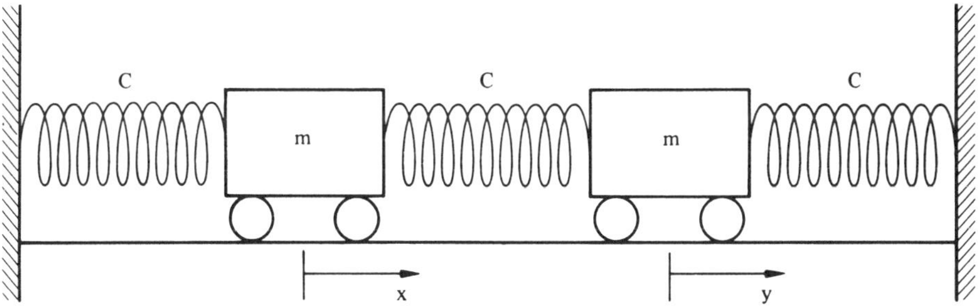

Fig. 4.2

4.4.3 Consider a body consisting of three point masses:

$m_1 = 1$ kg in $P_1(1/\sqrt{2}, 0, -1/\sqrt{2})$
$m_2 = 2$ kg in $P_2(1/2, 1/\sqrt{2}, 1/2)$
$m_3 = 5$ kg in $P_3(1/2, -1/\sqrt{2}, 1/2)$

- Determine its inertial tensor:

$$\mathbf{J} = \begin{bmatrix} \Sigma m\,(y^2 + z^2) & -\Sigma mxy & -\Sigma mxz \\ -\Sigma mxy & \Sigma m\,(x^2 + z^2) & -\Sigma myz \\ -\Sigma mxz & -\Sigma myz & \Sigma m\,(x^2 + y^2) \end{bmatrix}$$

- Calculate the largest principal moment of inertia and its associated axis using the iterated power algorithm, starting with

$$\mathbf{v} = \begin{bmatrix} 2 \\ -1.5 \\ 0 \end{bmatrix}$$

4.4.4 Use the iterated power algorithm to determine λ_3 and its associated normalized eigenvector $\mathbf{x}_3$ of the matrix S from the example in section 4.3.1.

Chapter 5

Polynomial Interpolation

5.1 INTRODUCTION

Let $y = f(x)$ be defined on some interval of the real axis, and let $x_0 < x_1 < \ldots < x_n$ be $n + 1$ distinct points of this interval. In the following, we assume that the values $y_i = f(x_i)$, $i = 0,1,\ldots, n$, of the function $y = f(x)$ are known (table 5.1).

Table 5.1

x	x_0	x_1	...	x_n
y	$f(x_0)$	$f(x_1)$	...	$f(x_n)$

When we wish to determine $f(x)$ in $\overline{x}$, where $\overline{x} \neq x_i$, $i = 0, 1,\ldots, n$, we speak of a problem of ***interpolation*** if $x_0 < \overline{x} < x_n$, and of a problem of ***extrapolation*** if $\overline{x} < x_0$, or $\overline{x} > x_n$.

These problems cannot be solved satisfactorily without additional information about $f(x)$. In practice, this information is often provided by the technical context.

Hereafter, we shall only address the problem of polynomial interpolation. To this end, we wish to approximate $f(x)$ using polynomials, although there are other classes of functions which can be used (for example, trigonometric, exponential, *et cetera*).

The polynomial $P_n(x)$ of degree n, whose graph passes through the $n + 1$ data points $(x_i, f(x_i))$, $i = 0, 1, \ldots, n$:

$$P_n(x) = a_n x^n + a_{n-1} x^{n-1} + \ldots + a_1 x + a_0$$

where **(5.1)**

$$P_n(x_i) = f(x_i), \qquad i = 0,1 \ldots, n,$$

is called the ***interpolating polynomial*** of $f(x)$ of degree n.

To determine $P_n(x)$, we may use many formulas, but, if the same data points $(x_i, f(x_i))$, $i=0, 1, \ldots, n$, are used, these formulas must all provide the same polynomial. Because, if $a_0, a_1, \ldots, a_n$ are the coefficients of an interpolating polynomial of degree n, it is necessary that

$$\begin{aligned} a_0 + a_1 x_0 + a_2 x_0^2 + \ldots + a_n x_0^n &= f(x_0) \\ a_0 + a_1 x_1 + a_2 x_1^2 + \ldots + a_n x_1^n &= f(x_1) \\ &\vdots \\ a_0 + a_1 x_n + a_2 x_n^2 + \ldots + a_n x_n^n &= f(x_n) \end{aligned} \tag{5.2}$$

This system of linear equations has exactly one solution $(a_0, a_1, \ldots, a_n)$, if its determinant:

$$V_{n+1} = \begin{vmatrix} 1 & x_0 & x_0^2 & \ldots & x_0^n \\ 1 & x_1 & x_1^2 & & x_1^n \\ \cdot & & & & \\ \cdot & & & & \\ \cdot & & & & \\ 1 & x_n & x_n^2 & & x_n^n \end{vmatrix}$$

which is called Vandermonde's determinant, does not vanish. Because $x_i \neq x_j$ for $i \neq j$, it can be easily shown that $V_{n+1} \neq 0$. Thus, for $n+1$ distinct interpolating points, there is one and only one interpolating polynomial of degree n.

Having determined the interpolating polynomial $P_n(x)$ of $f(x)$, we will be interested in an upper bound for $|f(\bar{x}) - P_n(\bar{x})|$, i.e., in the largest possible error introduced in approximating $f(x)$ by $P_n(x)$. It is from this point of view that the Lagrangian representation of the interpolating polynomial to be derived in the following section is more useful than system (5.2).

5.2 LINEAR INTERPOLATION

The simplest interpolation problem is as follows: knowing the two data points $(x_0, f(x_0))$, $(x_1, f(x_1))$, we approximate the value of $f(x)$, for any x between x_0 and x_1 by the straight line joining the two data points. We find the linear interpolation formula by superimposing the two linear polynomials $L_0(x)$ and $L_1(x)$ given by (see fig. 5.2)

$$L_0(x) = f(x_0)\,\frac{x - x_1}{x_0 - x_1}$$

$$L_1(x) = f(x_1)\,\frac{x - x_0}{x_1 - x_0}$$

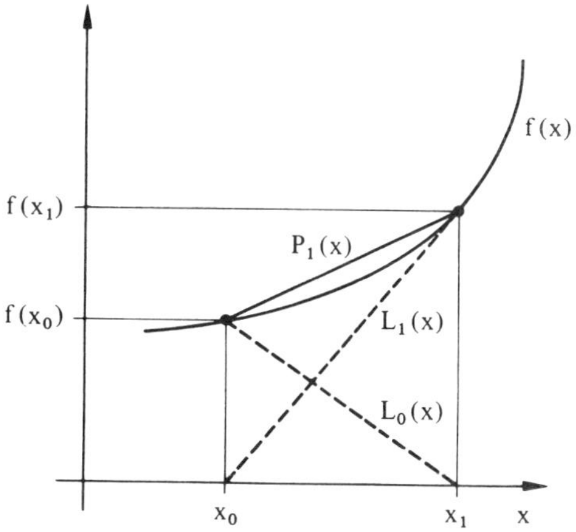

Fig. 5.2

The sum of L_0 and L_1 is a linear polynomial whose graph passes through the two data points. Thus, we find the ***first-order interpolating polynomial***:

$$P_1(x) = f(x_0)\,\frac{x - x_1}{x_0 - x_1} + f(x_1)\,\frac{x - x_0}{x_1 - x_0} \qquad \textbf{(5.3)}$$

5.3 QUADRATIC INTERPOLATION

Most data arise from graphs that are curved rather than straight. To better approximate such behavior, we look at polynomials of degree greater than one. Assume that three data points $(x_0, f(x_0))$, $(x_1, f(x_1))$ and $(x_2, f(x_2))$ of $f(x)$ are given. Then, we obtain the quadratic interpolation formula by superimposing the three quadratic polynomials (see fig. 5.3):

$$L_0(x) = f(x_0)\,\frac{(x-x_1)(x-x_2)}{(x_0-x_1)(x_0-x_2)}$$

$$L_1(x) = f(x_1)\,\frac{(x-x_0)(x-x_2)}{(x_1-x_0)(x_1-x_2)}$$

$$L_2(x) = f(x_2)\,\frac{(x-x_0)(x-x_1)}{(x_2-x_0)(x_2-x_1)}$$

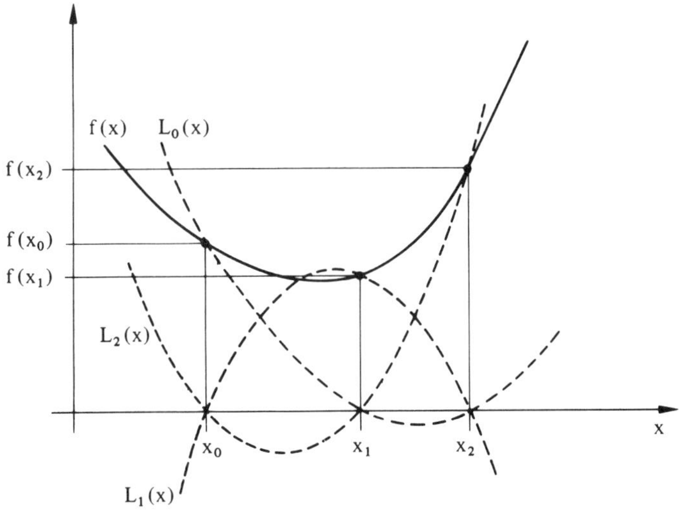

Fig. 5.3

Each of these three polynomials passes through one of the data points and vanishes at the two other interpolating points. The sum of the three polynomials:

$$P_2(x) = L_0(x) + L_1(x) + L_2(x)$$

is, therefore, the ***second-order interpolating polynomial*** which interpolates the three given data points. If the interpolating points x_i, $i = 0, 1, 2$, are equidistant, we have

$$P_2(x) = f(x_1) + \frac{f(x_2)-f(x_0)}{2}\,\frac{x-x_1}{\Delta x} + \frac{f(x_0)-2f(x_1)+f(x_2)}{2}\,\frac{(x-x_1)^2}{\Delta x^2}$$

where **(5.4)**

$$\Delta x = x_2 - x_1 = x_1 - x_0$$

5.4 LAGRANGE'S INTERPOLATION FORMULA

The two formulas (5.3) and (5.4) are special cases of Lagrange's interpolation formula. To obtain Lagrange's formula, it is assumed that we know the $n+1$ data points $(x_i, f(x_i))$, $i=0, 1, \dots, n$, of the function $y=f(x)$; the interpolating points x_i are all distinct and may or may not be equidistant.

To obtain the desired interpolation formula, we consider the ***Lagrange polynomials***:

$$L_k(x)=f(x_k)\prod_{\substack{i=0\\ i\neq k}}^{n}\frac{(x-x_i)}{(x_k-x_i)}, \qquad k=0, 1, \dots, n, \qquad \mathbf{(5.5)}$$

i.e., the polynomials of degree n which take the value $f(x_k)$ for $x=x_k$ and 0 for $x=x_i$, $i\neq k$. The ***Lagrangian interpolating formula*** is given by the sum of the n polynomials (5.5):

$$P_n(x)=\sum_{k=0}^{n} L_k(x) \qquad (5.6)$$

If $E_n(x)=f(x)-P_n(x)$ designates the error introduced when we substitute the unknown function $f(x)$ by the polynomials $P_n(x)$, it can be demonstrated using Rolle's theorem that

$$E_n(x)=(x-x_0)(x-x_1)\dots(x-x_n)\,\frac{f^{(n+1)}(\xi)}{(n+1)!}$$

where $\xi\in(x_0, x_n)$, assuming that $f(x)$ has continuous derivatives up to the order $n+1$.

The Lagrangian interpolating formula is especially useful if the interpolating points remain the same, but the interpolated values change. The formula is less suitable for a change of the interpolating points or a change in the number of data points. On the other hand, it allows us to find numerical integration and differentiation formulas easily (see sections 5.5 and 5.6).

5.4.1 Example: Cubic interpolation

Let us determine the Lagrangian interpolating polynomial which satisfies table 5.4. The degree of the polynomial is three (it is always equal to the number of interpolating points minus one).

Table 5.4

x	0	2	3	5
y	−1	2	9	87

We calculate

$$L_0(x) = f(x_0)\ \frac{(x-x_1)(x-x_2)(x-x_3)}{(x_0-x_1)(x_0-x_2)(x_0-x_3)} = -1\ \frac{(x-2)(x-3)(x-5)}{(-2)(-3)(-5)}$$

$$= \frac{1}{30}\ (x^3 - 10x^2 + 31x - 30)$$

$$L_1(x) = f(x_1)\ \frac{(x-x_0)(x-x_2)(x-x_3)}{(x_1-x_0)(x_1-x_2)(x_1-x_3)} = 2\ \frac{x(x-3)(x-5)}{2(-1)(-3)}$$

$$= \frac{1}{3}\ (x^3 - 8x^2 + 15x)$$

$$L_2(x) = f(x_2)\ \frac{(x-x_0)(x-x_1)(x-x_3)}{(x_2-x_0)(x_2-x_1)(x_2-x_3)} = 9\ \frac{x(x-2)(x-5)}{(3)(1)(-2)}$$

$$= -\frac{3}{2}\ (x^3 - 7x^2 + 10x)$$

$$L_3(x) = f(x_3)\ \frac{(x-x_0)(x-x_1)(x-x_2)}{(x_3-x_0)(x_3-x_1)(x_3-x_2)} = 87\ \frac{x(x-2)(x-3)}{5 \cdot 3 \cdot 2}$$

$$= \frac{29}{10}\ (x^3 - 5x^2 + 6x)$$

We find

$$P_3(x) = L_0(x) + L_1(x) + L_2(x) + L_3(x) = \frac{53}{30}\ x^3 - 7x^2 + \frac{253}{30}\ x - 1.$$

5.5 NUMERICAL INTEGRATION

The central idea behind numerical integration is to replace the function in question by an approximating function whose integral can be evaluated analytically. In this context, we have just approximated a function $f(x)$ by using interpolation polynomials. By integrating these polynomials, we may find approximations of the integral of $f(x)$.

For simplicity, we shall use the notation $x_i = x_0 + i\Delta x$, where $\Delta x = x_{i+1} - x_i = \text{const.}$, $i = 0, 1, \ldots, n$; $f_i = f(x_i)$.

5.5.1 Trapezoidal rule

To obtain the trapezoidal rule, we first approximate the given function $f(x)$ in the interval $[x_0, x_1]$ by the line joining the two data points (x_0, f_0) and (x_1, f_1) (see (5.3)), obtaining for the interpolating linear polynomial:

$$f(x) \approx f_0 + \frac{x - x_0}{\Delta x}(f_1 - f_0)$$

Thus we obtain for the integral:

$$\int_{x_0}^{x_1} f(x)\,dx \approx \frac{1}{2}\Delta x(f_0 + f_1) \tag{5.7}$$

and by repeating this procedure for all the subintervals under consideration, we find the ***trapezoidal numerical integration rule***

$$\int_{x_0}^{x_n} f(x)\,dx \approx \frac{1}{2}\Delta x(f_0 + 2f_1 + \ldots + 2f_{n-1} + f_n) \tag{5.8}$$

If $f(x)$ can be expanded in a Taylor series, the accuracy of the trapezoidal rule (5.8) can be improved in the following way. We develop $f(x)$ about the point x_0:

$$f(x) = f_0 + (x - x_0)\,f_0' + \frac{(x - x_0)^2}{2!}\,f_0'' + \ldots$$

where $f_0^{(k)} = f^{(k)}(x_0)$, and by integrating $f(x)$ between the limits x_0 and x_1, we obtain

$$\int_{x_0}^{x_1} f(x)\,dx = \Delta x f_0 + \frac{1}{2}\Delta x^2 f_0' + \frac{1}{6}\Delta x^3 f_0'' + \ldots \tag{5.9}$$

Since

$$f(x_1) = f_1 = f_0 + \Delta x f_0' + \frac{1}{2}\Delta x^2 f_0'' + \ldots$$

we have

$$\frac{1}{2}\,\Delta x f_0 = \frac{1}{2}\,\Delta x f_1 - \frac{1}{2}\,\Delta x^2 f_0' - \frac{1}{4}\,\Delta x^3 f_0'' - \ldots \tag{5.10}$$

If we substitute half of the first term on the right-hand side of (5.9) by the right side of (5.10), we find, for the integral over a single subinterval of length Δx:

$$\int_{x_0}^{x_1} f(x)\,dx = \frac{1}{2}\,\Delta x\,(f_0 + f_1) - \frac{1}{12}\,\Delta x^3 f_0'' + \ldots$$

and over the n subintervals from x_0 to x_n:

$$\int_{x_0}^{x_n} f(x)\,dx = \frac{1}{2}\,\Delta x\,(f_0 + 2f_1 + \ldots + 2f_{n-1} + f_n) - \frac{1}{12}\,\Delta x^2 \sum_{k=0}^{n-1} f_k''\,\Delta x + \ldots$$

In this last formula, the sum:

$$\sum_{k=0}^{n-1} f_k''\,\Delta x$$

equals the area of the n rectangles of heights f_k'', $k = 0, 1, \ldots, n-1$, and of width Δx. This sum can be approximated by the integral of the second derivative f'' over the interval $[x_0, x_n]$. Thus, we find the ***improved trapezoidal integration rule***:

$$\int_{x_0}^{x_n} f(x)\,dx \approx \frac{1}{2}\,\Delta x\,(f_0 + 2f_1 + \ldots + 2f_{n-1} + f_n) - \frac{1}{12}\,\Delta x^2 (f_n' - f_0') \tag{5.11}$$

5.5.2 Simpson's rule

To improve on the trapezoidal integration rule, we use quadratic interpolation, obtaining Simpson's rule. To this end, we first approximate $f(x)$ in the interval $[x_0, x_2]$ by the parabola which passes through the three points (x_0, f_0), (x_1, f_1), and (x_2, f_2), obtaining the quadratic interpolating formula (see (5.4)):

$$f(x) \approx f_1 + \frac{x - x_1}{\Delta x}\,\frac{f_2 - f_0}{2} + \frac{(x - x_1)^2}{\Delta x^2}\,\frac{f_0 - 2f_1 + f_2}{2}$$

Thus, we obtain for the integral

$$\int_{x_0}^{x_2} f(x)\,dx \approx \frac{1}{3}\,\Delta x\,(f_0 + 4f_1 + f_2)$$

and by repeating this procedure for an ***even*** number of subintervals, we find ***Simpson's rule***, which is the most popular numerical integration method:

$$\int_{x_0}^{x_n} f(x)\,dx \approx \frac{1}{3}\,\Delta x\,(f_0 + 4f_1 + 2f_2 + \ldots + 2f_{n-2} + 4f_{n-1} + f_n) \tag{5.12}$$

If f(x) can be expanded in a Taylor series, we can improve formula (5.12) in the same way as we have just improved the trapezoidal formula (5.10), by developing f(x) about the point x_1:

$$f(x) = f_1 + (x - x_1)\,f_1' + \frac{(x - x_1)^2}{2\,!}\,f_1'' + \frac{(x - x_1)^3}{3\,!}\,f_1''' + \frac{(x - x_1)^4}{4\,!}\,f_1^{IV} + \ldots$$

and since $x_2 = x_1 + \Delta x$, $x_0 = x_1 - \Delta x$, we calculate

$$\int_{x_0}^{x_2} f(x)\,dx = 2\,\Delta x\,f_1 + \frac{1}{3}\,\Delta x^3\,f_1'' + \frac{1}{60}\,\Delta x^5\,f_1^{IV} + \ldots \tag{5.13}$$

where the even power terms cancel out. We then evaluate f(x) at the points x_0 and x_2:

$$f_0 = f_1 - \Delta x\,f_1' + \frac{1}{2\,!}\,\Delta x^2 f_1'' - \frac{1}{3\,!}\,\Delta x^3\,f_1''' + \frac{1}{4\,!}\,\Delta x^4\,f_1^{IV} - \ldots$$

$$f_2 = f_1 + \Delta x\,f_1' + \frac{1}{2\,!}\,\Delta x^2 f_1'' + \frac{1}{3\,!}\,\Delta x^3\,f_1''' + \frac{1}{4\,!}\,\Delta x^4\,f_1^{IV} + \ldots$$

and we find, after an algebraic transformation,

$$2\,\Delta x\,f_1 = \frac{1}{3}\,\Delta x\,(f_0 + 4f_1 + f_2) - \frac{1}{3}\,\Delta x^3 f_1'' - \frac{1}{36}\,\Delta x^5 f_1^{IV} - \ldots \tag{5.14}$$

By substituting the first term on the right-hand side of (5.13) by (5.14), we thus obtain

$$\int_{x_0}^{x_2} f(x)\,dx = \frac{1}{3}\Delta x\,(f_0 + 4f_1 + f_2) - \frac{1}{90}\,\Delta x^5 f_1^{IV} + \ldots$$

If we integrate f(x) over the interval $[x_0, x_n]$, always assuming that n is even, we find

$$\int_{x_0}^{x_n} f(x)\,dx = \frac{1}{3}\Delta x(f_0 + 4f_1 + 2f_2 + \ldots + 2f_{n-2} + 4f_{n-1} + f_n)$$

$$-\frac{1}{180}\Delta x^4 \sum_{k=1}^{n/2} f_{2k-1}^{IV}\, 2\Delta x + \ldots$$

Because

$$\sum_{k=1}^{n/2} f_{2k-1}^{IV}\, 2\Delta x \approx \int_{x_0}^{x_n} f^{IV}\,dx = f_n''' - f_0'''$$

we have the ***improved Simpson's formula***:

$$\int_{x_0}^{x_n} f(x)\,dx \approx \frac{1}{3}\Delta x(f_0 + 4f_1 + 2f_2 + \ldots + 2f_{n-2} + 4f_{n-1} + f_n)$$

$$-\frac{1}{180}\Delta x^4 (f_n''' - f_0''') \qquad \textbf{(5.15)}$$

5.5.3 Note on the error

When the trapezoidal rule (5.8) is used, the error is proportional to Δx^2 (see (5.11)), and when Simpson's rule (5.12) is used, the error is proportional to Δx^4 (see (5.15)). The integration interval can always be shifted into the interval $0 \leq x \leq 1$ by means of a linear transformation of the integration variable. In this case, we have $\Delta x < 1$, and thus $\Delta x^4 < \Delta x^2$, i.e., Simpson's rule provides a better approximation of the integral than the trapezoidal formula, except when the first derivative f′ is a very small or the third derivative f‴ is very large.

5.5.4 Example: Comparison of the two numerical integration methods

We wish to determine

$$\int_0^{\pi/2} f(x)\,dx$$

on the basis of table 5.5

Table 5.5

x	0	$\pi/8$	$\pi/4$	$3\pi/8$	$\pi/2$
f(x)	0	0.382683	0.707107	0.923880	1

- According to the trapezoidal rule, we find

$$\int_0^{\pi/2} f(x)dx \approx \frac{1}{2}\Delta x(f_0 + 2f_1 + 2f_2 + 2f_3 + f_4)$$

$$= \frac{1}{2}\frac{\pi}{8}(0 + 2\cdot 0.382683 + 2\cdot 0.707107$$

$$+\ 2\cdot 0.923880 + 1.000000) = 0.987116$$

- According to Simpson's rule, we find

$$\int_0^{\pi/2} f(x)dx \approx \frac{1}{3}\Delta x(f_0 + 4f_1 + 2f_2 + 4f_3 + f_4)$$

$$= \frac{1}{3}\frac{\pi}{8}(0 + 4\cdot 0.382683 + 2\cdot 0.707107$$

$$+\ 4\cdot 0.923880 + 1.000000) = 1.000135$$

The data points given are the data points of the function $f(x) = \sin x$; thus, we have

$$\int_0^{\pi/2} \sin x\, dx = 1$$

We find that the approximation by Simpson's rule is clearly better than that provided by the trapezoidal rule.

5.6 NUMERICAL DIFFERENTIATION

The derivative $f'(x)$ of a function $f(x)$ can be approximated by the derivative $P_n'(x)$ of the interpolating polynomial $P_n(x)$ of $f(x)$. Because this is a double approximation, we can expect precision problems. Figure 5.6 illustrates the situation.

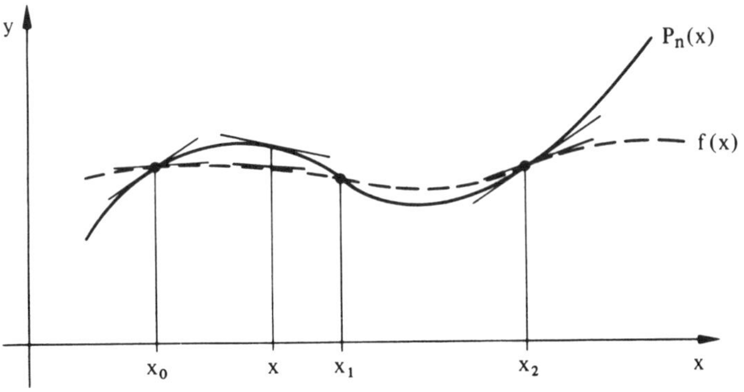

Fig. 5.6

Assuming that the interpolating points are equidistant, we obtain, for example, by differentiating the interpolating polynomial $P_2(x)$ (see (5.4)), for the first derivative $f'(x)$:

$$f'(x) \approx P_2'(x) = \frac{f_2 - f_0}{2\Delta x} + (f_0 - 2f_1 + f_2)\,\frac{x - x_1}{\Delta x^2} \tag{5.16}$$

and for the second derivative $f''(x)$:

$$f''(x) \approx P_2''(x) = \frac{f_0 - 2f_1 + f_2}{\Delta x^2} \tag{5.17}$$

5.6.1 Application to the solution of Laplace's equation

We wish to solve ***Laplace's equation*** numerically, i.e., to make an approximation of the function $u(x,y)$ which satisfies the partial differential equation:

$$\Delta u = \frac{\partial^2 u}{\partial x^2} + \frac{\partial^2 u}{\partial y^2} = 0 \tag{5.18}$$

in a bounded domain D with boundary C, assuming that we know the values of u on C (Dirichlet's problem).

To replace the Laplace equation by a finite difference approximation, we impose a square grid on D of horizontal and vertical separation h, such that C passes through the points of the grid (fig. 5.7).

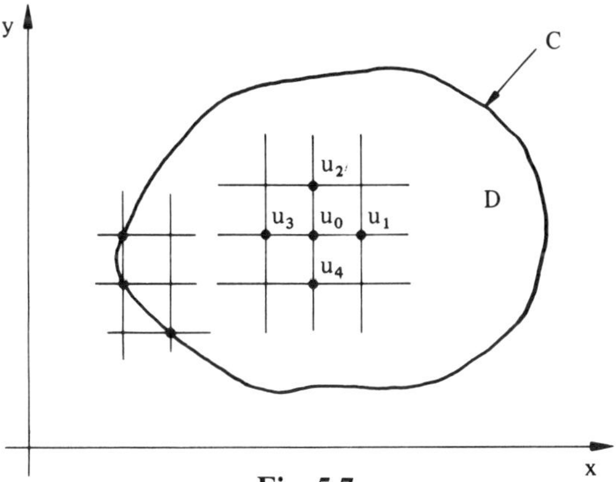

Fig. 5.7

The partial derivatives are approximated at a point (x_0, y_0) of the grid in the interior of D in the following manner. Designating the values of u at $(x_0 \pm h, y_0 \pm h)$ by u_i, i = 1, 2, 3, 4, we find, from (5.17)

$$\left.\frac{\partial^2 u}{\partial x^2}\right|_0 \approx \frac{u_1 - 2u_0 + u_3}{h^2} \qquad \left.\frac{\partial^2 u}{\partial y^2}\right|_0 \approx \frac{u_2 - 2u_0 + u_4}{h^2}$$

from which we get

$$\Delta u|_0 \approx \frac{u_1 + u_2 + u_3 + u_4 - 4u_0}{h^2} \tag{5.19}$$

According to (5.18), we thus have

$$0 \approx u_1 + u_2 + u_3 + u_4 - 4u_0$$

or

$$u_0 \approx \frac{1}{4}(u_1 + u_2 + u_3 + u_4) \tag{5.20}$$

Formula (5.20) is called the ***five-point formula***.

For each point of the grid inside D, we have the five-point formula (5.20). Thus, if there are n such points, we obtain n linear equations to approximate the values of u at these points.

An estimate for the approximation error of the Laplace operator by the numerical operator (5.19) can be found by expanding all the terms on the right side of (5.19) in a Taylor series about the point (x_0, y_0). We obtain for terms of order four:

$$u_1 + u_2 + u_3 + u_4 - 4u_0 = h^2\left(\frac{\partial^2 u}{\partial x^2} + \frac{\partial^2 u}{\partial y^2}\right) - \frac{1}{3!} h^4 \frac{\partial^4 u}{\partial x^2 \partial y^2}$$

thus,

$$\epsilon = \Delta u - \frac{u_1 + u_2 + u_3 + u_4 - 4u_0}{h^2} = \frac{1}{6} h^2 \frac{\partial^4 u}{\partial x^2 \partial y^2}$$

Finally, we note that the solution of ***Poisson's equation*** can be approximated analogously

$$\Delta u = \frac{\partial^2 u}{\partial x^2} + \frac{\partial^2 u}{\partial y^2} = f(x,y) \qquad \textbf{(5.21)}$$

In this case, the five-point formula takes the following form:

$$u_0 \approx \frac{1}{4}(u_1 + u_2 + u_3 + u_4) - \frac{1}{4} h^2 f_0 \qquad \textbf{(5.22)}$$

where $f_0 = f(x_0, y_0)$.

5.7 EXERCISES

5.7.1 Given $f(x) = 1/(1 + x^2)$,

- determine the Lagrangian interpolating polynomial for the interpolating points –2, –1, 0, 1, 2;
- discuss the interpolation error.

5.7.2 Give a geometric illustration of the trapezoidal rule.

5.7.3 A rocket is fired vertically from the ground, and its acceleration during the first 80 seconds is measured (table 5.8). Using Simpson's rule, determine

- the speed of the rocket at the time $t = 80$ sec;
- the height of the rocket at time $t = 80$ sec.

Table 5.8

t(sec)	0	10	20	30	40	50	60	70	80
$a(m/sec^2)$	30.00	31.63	33.44	35.47	37.75	40.33	43.29	46.70	50.67

5.7.4 Calculate numerically the integral:

$$\int_0^1 \exp(-x^2)\, dx$$

using the following formulas:

- trapezoidal rule,
- improved trapezoidal rule,
- Simpson's rule,

setting $\Delta x = 0.5$, $\Delta x = 0.25$, $\Delta x = 0.125$, and $\Delta x = 0.0625$,

- compare the results.

5.7.5 Calculate the integral (see fig. 5.9):

$$\iint_D x\, y\, dx\, dy$$

- analytically,
- using Simpson's rule, setting $\Delta x = \Delta y = 1$.

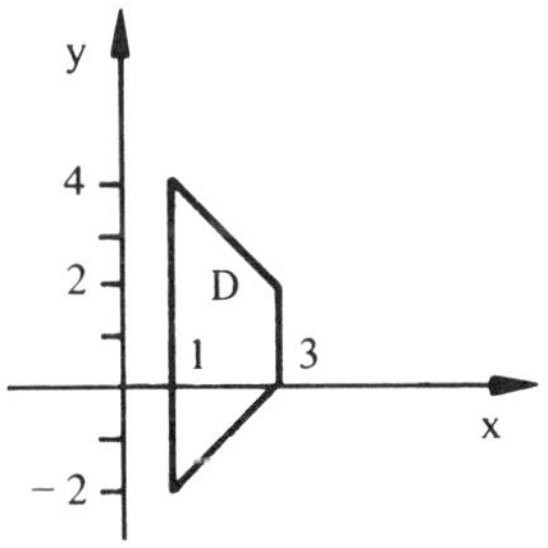

Fig. 5.9

5.7.6 Given the boundary value problem depicted in figure 5.10:

- find the approximations u_k, $k = 1, 2, 3, 4$, of the solution of Laplace's equation $\Delta u = 0$ using the five-point formula. The boundary values of u are indicated in the figure;

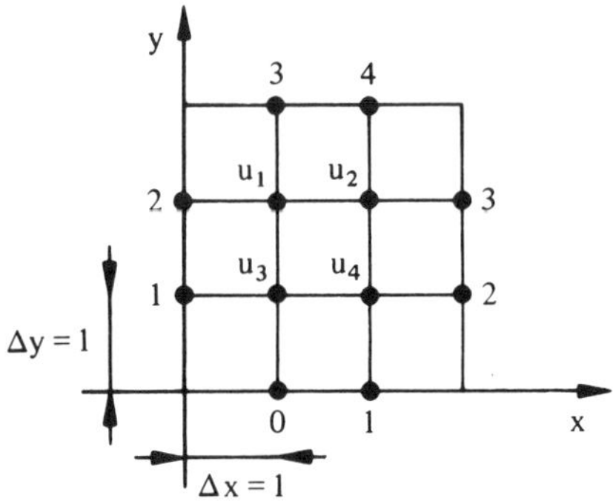

Fig. 5.10

- carry out three iterations of the Jacobi method to solve the linear equations obtained.

5.7.7 Given the boundary value problem of figure 5.11, find the approximations u_k, $k = 1, 2, 3$, of the solution of Poisson's equation $\Delta u = xy$ using the five-point formula. The boundary values of u are indicated in the figure.

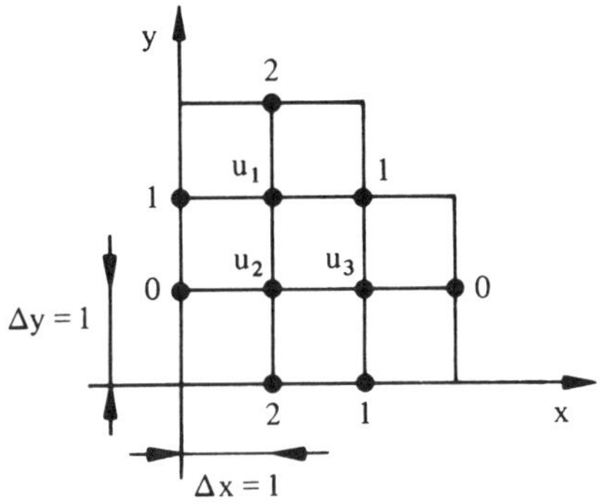

Fig. 5.11

Chapter 6

Solution of Differential Equations by Graphical and Numerical Methods

The number of differential equations permitting an analytical solution is relatively small; numerical and graphical methods, therefore, are of great practical interest.

6.1 GRAPHICAL METHOD OF ISOCLINES APPLIED TO THE FIRST-ORDER EQUATION

We wish to find the function y (x) satisfying the first-order differential equation:

$$y' = F(x, y) \qquad (6.1)$$

subject to the initial condition $y(x_0) = y_0$.

Equation (6.1) states that at any point (x,y) the slope of the curve y = y (x) equals F (x,y). The method of ***isoclines*** is based on the family of curves F (x,y) = constant, which are not themselves solutions, but are helpful in determining the character of solutions. Where a solution y = y (x) of the differential equation (6.1) crosses one of the isoclines F (x,y) = c, it must have for its slope the constant c of that isocline. To satisfy the initial condition $y(x_0) = y_0$, we shall determine the curve y = y (x) passing through the given initial point (x_0, y_0) (see fig. 6.1).

6.1.1 Example: Solution of a first-order equation

Given the equation:

$$y' = x^2 + y^2$$

The isoclines are the circles $x^2 + y^2 = r^2 = c$. Figure 6.2 shows the construction of the trajectories corresponding to the initial conditions $(x_0, y_0) = (0,0)$ and $(x_0^*, y_0^*) = (0,1)$.

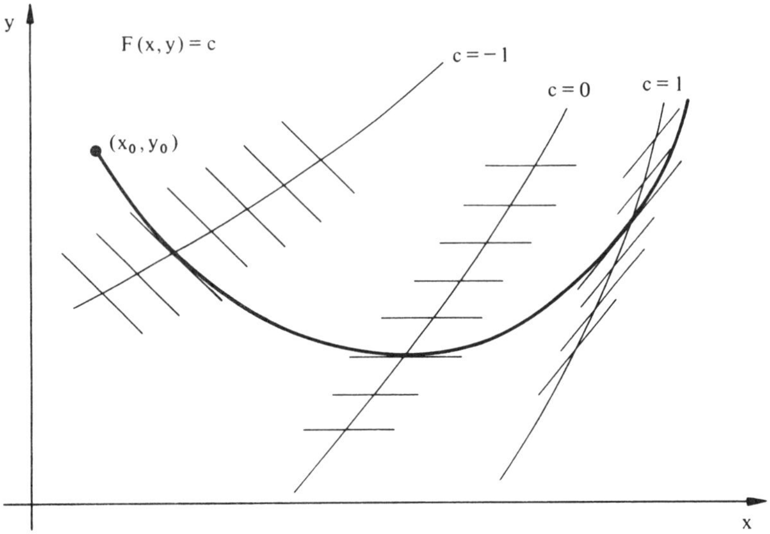

Fig. 6.1

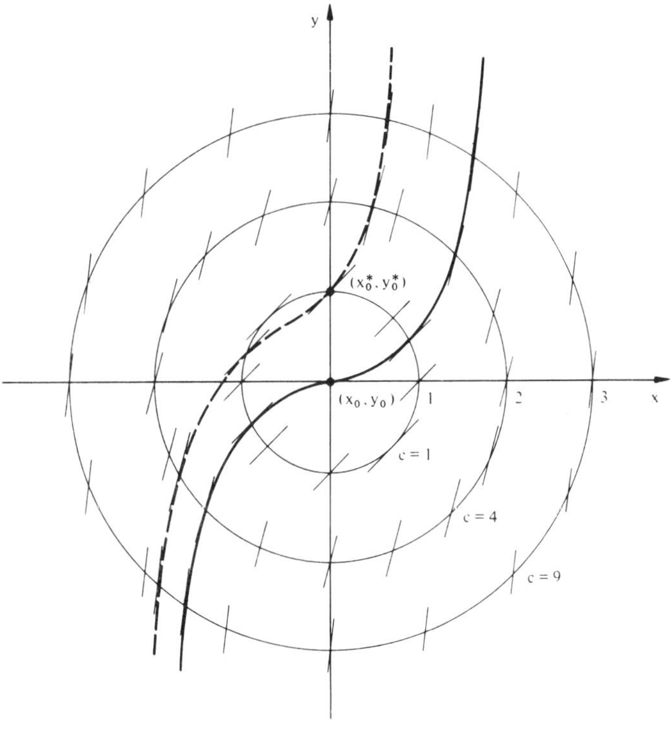

Fig. 6.2

6.2 GRAPHICAL METHOD OF ISOCLINES APPLIED TO AUTONOMOUS SECOND-ORDER EQUATIONS

We wish to find the function x(t) satisfying the second-order differential equation:

$$\ddot{x} = F(x, \dot{x}, t) \tag{6.2}$$

where the dot indicates the derivative with respect to t.

Equation (6.2) is called ***autonomous***, if

$$\ddot{x} = F(x, \dot{x}) \tag{6.3}$$

i.e., if F does not depend explicitly on t.

The equations of type (6.2) are generally difficult to solve explicitly. Nonetheless, in the case of autonomous equations (6.3), we can deduce the important characteristics of the solutions in the $(x, \dot{x})$-plane which is called the ***phase plane***. In this case, we express $\dot{x}$ as a function of x by setting:

$$\dot{x} = y(x)$$

from which we get

$$\ddot{x} = y'(x)\,\dot{x} = y'y$$

Substituting these expressions into (6.3), we obtain

$$y'y = F(x, y) \tag{6.4}$$

which constitutes a first-order differential equation relating $y = \dot{x}$ and x.

A pair of values $(x, \dot{x})$ is called a ***system state***.

6.2.1 Example: Solution of a second-order autonomous equation

Given the autonomous differential equation:

$$\ddot{x} + \dot{x} + x = 0 \tag{a}$$

Setting $\dot{x} = y(x)$, we have

$$\ddot{x} = y'(x)\,\dot{x} = y'y$$

and equation (a) becomes

$$y'y + y + x = 0$$

or

$$y' = -1 - \frac{x}{y} \tag{b}$$

We wish to use the isocline method to determine the solutions of equation (b) in the phase plane x, $\dot{x} = y$. The isoclines $y' = c$ are given by the lines $y = -x/(c+1)$, and figure 6.3 shows the construction of the trajectories that are always traversed toward the origin of the phase plane. Because $\dot{x} > 0$ in the upper half-plane, x increases in the same half-plane, and because $\dot{x} < 0$ in the lower half-plane, x decreases in the same half-plane.

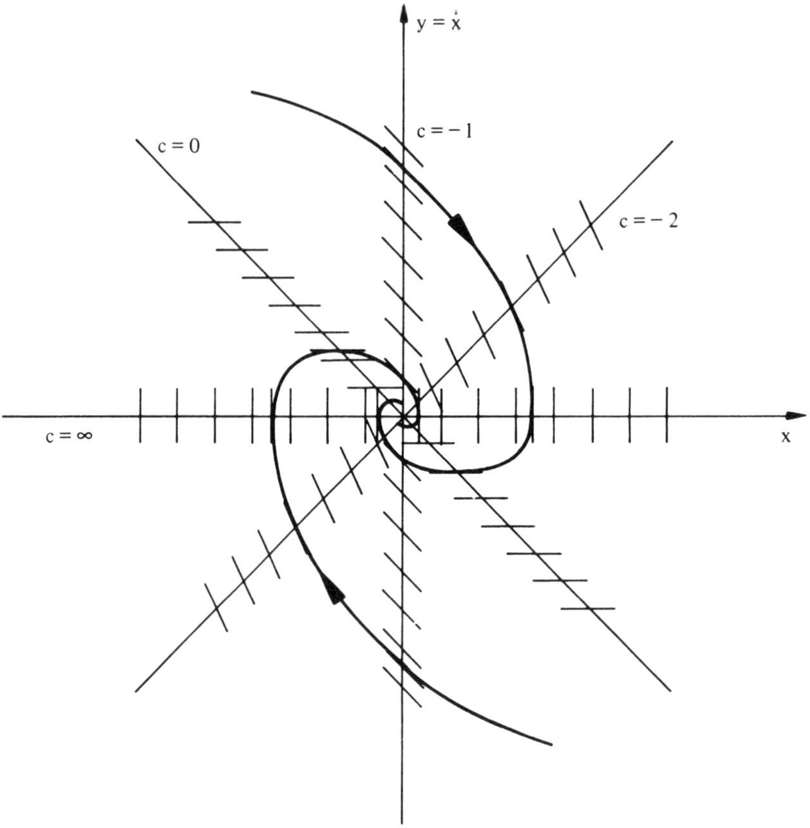

Fig. 6.3

6.2.2 Example: Movement of the pendulum

Given the differential equation of motion of the pendulum (see fig. 6.4)

$$a\ddot{x} + g \sin x = 0 \tag{a}$$

Setting $\dot{x} = y(x)$, we have

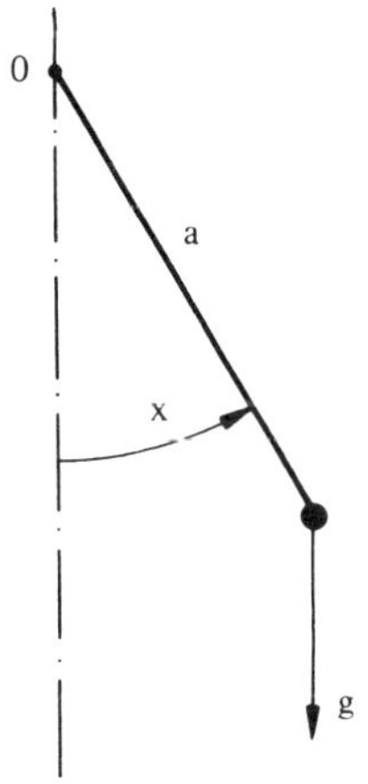

Fig. 6.4

$$\ddot{x} = y' \dot{x} = y' y$$

and equation (a) becomes

$$a\, y' y + g \sin x = 0 \qquad \text{(b)}$$

Before applying the isocline method to the first-order equation (b), we wish to solve this equation analytically. To this end, we first write it in the form

$$a\, y\, dy + g \sin x\, dx = 0$$

Then, integrating, we obtain

$$a \frac{y^2}{2} - g \cos x = c$$

or again, as $\dot{x} = y$:

$$\dot{x} = \sqrt{\frac{2}{a}(c + g \cos x)} \qquad \text{(c)}$$

Equation (c) relates $\dot{x}$ and x instead of x and t, and figure 6.5 shows the trajectories in the phase plane x, $\dot{x}$ for different values of c. To study the evolution of the system in time, we note that x is increasing in the upper half-plane, since $\dot{x} > 0$ in this half-plane. Consequently, the trajectories are traversed from left to right. In an analogous fashion, we see that they are traversed from right to left in the lower half-plane.

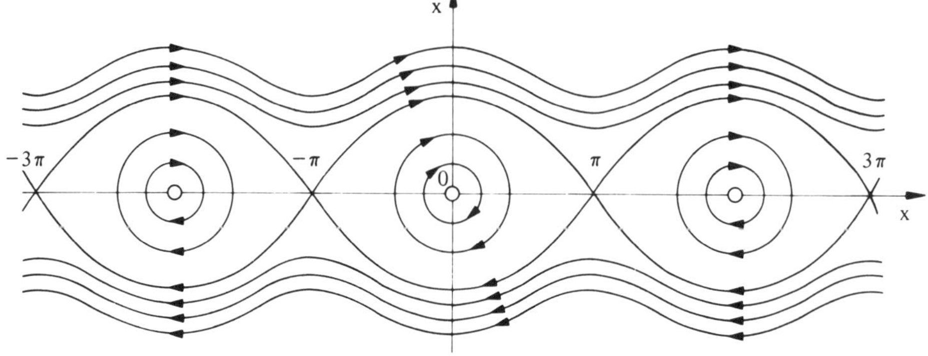

Fig. 6.5

Finally, in order to apply the isocline method to the solution of the problem, we set $y' = c$ in equation (b), from which we get the isocline equation:

$$y = -\frac{g \sin x}{ac}$$

Figure 6.6 illustrates the graphical construction of the trajectories of equation (b) for $g/a = 1$.

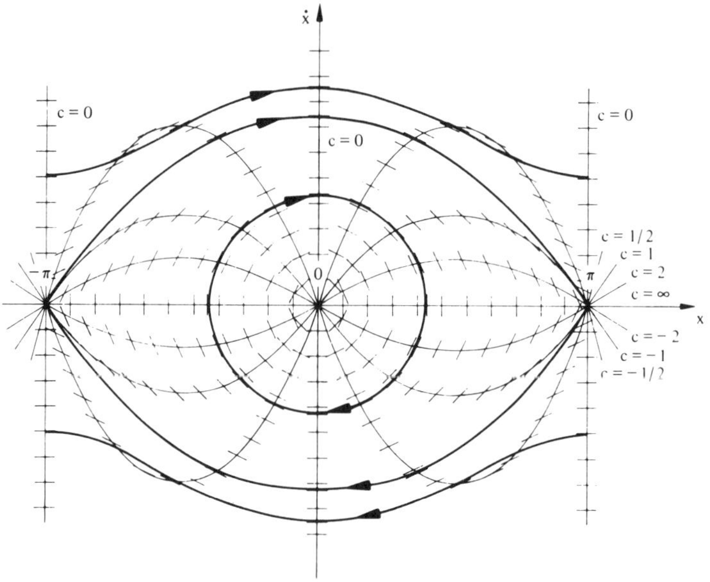

Fig. 6.6

6.3 NUMERICAL INTEGRATION METHODS FOR THE FIRST-ORDER EQUATION

Let y (x) be the solution of the first-order differential equation:

$$y' = F(x, y), \qquad y(x_0) = y_0 \tag{6.5}$$

where F (x,y) is defined in a domain of the xy-plane.

The problem consists of approximating the solution y (x) of (6.5) numerically in the interval $[x_0, b]$, subject to the initial condition $y(x_0) = y_0$.

In all the numerical methods developed hereafter, the interval $[x_0, b]$ is subdivided into n subintervals of length $\Delta x = (b - x_0)/n$ obtaining the discrete set of evenly spaced nodes $x_i = x_0 + i\Delta x, i = 0, 1, \ldots, n$.

6.3.1 Euler's method

At x_0, we know y_0, and thus $y'(x_0) = F(x_0, y_0)$ is also known. According to Euler's method, the solution of equation (6.5) on the subinterval $[x_0, x_1]$ is approximated by its tangent at x_0. Thus, we have

$$y_1 = y_0 + \Delta x\, F(x_0, y_0)$$

In the subinterval $[x_1, x_2]$, the solution of (6.5) will be replaced by the tangent at the point (x_1, y_1). We find

$$y_2 = y_1 + \Delta x\, F(x_1, y_1)$$

and so on. ***Euler's algorithm*** is thus given by (see fig. 6.7):

$$y_{i+1} = y_i + \Delta x\, F(x_i, y_i), \qquad i = 0, 1, \ldots, n-1 \tag{6.6}$$

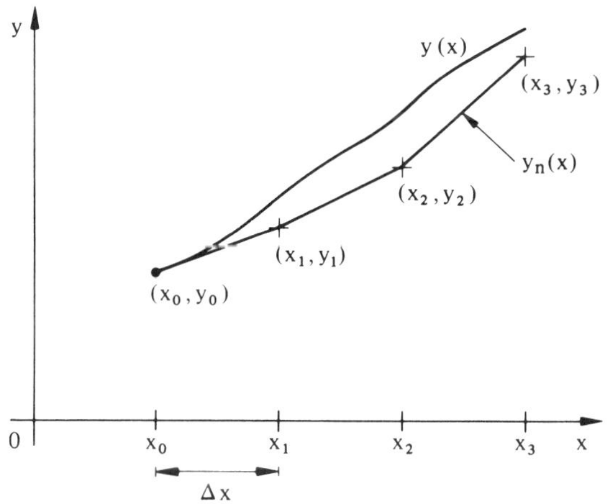

Fig. 6.7

Euler's method is not an efficient numerical method, but many of the ideas involved in the numerical solution of differential equations are introduced most simply with it.

Regarding the convergence of this method, we mention the following fact without demonstration: let $y = y_n(x)$ be the resulting polygon chains for the sequence of decreasing integrating steps $\Delta x = (b - x_0)/n$, $n = 1, 2, 3, \ldots$ If there is a unique solution $y = y(x)$ of (6.5) on $[x_0, b]$, subject to $y(x_0) = y_0$, then $|y_n(x) - y(x)|$ converges to 0 as n approaches $+\infty$, for all $x \epsilon [x_0, b]$.

6.3.2 Example: Solution of an equation using Euler's method

Given the equation:

$$y' = y + x, \qquad y(0) = 1 \tag{a}$$

We wish to approximate the solution of (a) at $x = 1$, using Euler's method, by subdividing the segment [0,1] into ten equal subintervals.

According to formula (6.6),

$$y_{i+1} = y_i + \Delta x\,(y_i + x_i), \qquad i = 0, 1, \ldots, 9; \qquad \Delta x = 0.1$$

we calculate the values of table 6.8.

Table 6.8

i	0	1	2	3	4	5	6	7	8	9	10
x_i	0	0.1	0.2	0.3	0.4	0.5	0.6	0.7	0.8	0.9	1.0
y_i	1	1.1	1.22	1.362	1.524	1.7164	1.9380	2.1918	2.4810	2.8091	3.1800

We find $y(1) \approx 3.1800$. The exact solution of equation (a) is given by $y = 2\exp(x) - x - 1$, which gives $y(1) = 3.4366$.

The numerical approximation is thus very inaccurate. It can be improved by choosing a smaller integration step Δx, or by considering higher-order Taylor approximations, as will be done in the following section.

6.3.3 Taylor integration method

Instead of using straight-line segments to approximate the solution of equation (6.5), we will use arcs of parabolas.

According to Taylor's formula, we have

$$y_{i+1} = y(x_{i+1}) = y(x_i + \Delta x)$$

$$= y(x_i) + \Delta x\, y'(x_i) + \frac{1}{2!} \Delta x^2\, y''(x_i) + \dots$$

and since

$$y''(x_i) = \frac{d}{dx}(y'(x))_{x=x_i} = \frac{d}{dx}(F(x, y))_{x=x_i}$$

$$= \left(\frac{\partial F}{\partial x} + \frac{\partial F}{\partial y} \frac{dy}{dx} \right)_{x=x_i} = \left(\frac{\partial F}{\partial x} + \frac{\partial F}{\partial y} y' \right)_{x=x_i}$$

$$= \frac{\partial F}{\partial x}(x_i, y_i) + \frac{\partial F}{\partial y}(x_i, y_i)\, F(x_i, y_i)$$

we find, leaving out the higher-order terms, the following algorithm:

$$y_{i+1} = y_i + \Delta x\, F(x_i, y_i) + \frac{1}{2} \Delta x^2 \left(\frac{\partial F}{\partial x} + \frac{\partial F}{\partial y} F \right)_{(x_i,\, y_i)} \qquad \textbf{(6.7)}$$

$i = 0, 1, \dots, n - 1$

It can be easily shown that algorithm (6.7) is more accurate than that of Euler (6.6) for the same integration step. If we still wish to improve the speed of convergence, we decrease the step size or use a larger number of terms in the Taylor series, and consequently adapt formula (6.7).

6.3.4 Example: Solution of an equation using the Taylor method

For equation (b) of section 6.2.2. (movement of the pendulum):

$$y' = -\frac{g}{a} \frac{\sin x}{y} = F(x, y)$$

we obtain, from (6.7), the following algorithm:

$$x_{i+1} = x_i + \Delta x$$

$$y_{i+1} = \dot{x}_{i+1} = y_i - \Delta x \frac{g}{a} \frac{\sin x_i}{y_i} + \frac{1}{2} \Delta x^2 \left(-\frac{g}{a} \frac{\cos x_i}{y_i} - \frac{g^2}{a^2} \frac{\sin^2 x_i}{y_i^3} \right)$$

$$i = 0, 1, \ldots, n-1$$

6.3.5 Runge-Kutta methods

The Taylor integration method is conceptually easy to work with, but it is time-consuming to have to calculate the higher-order derivatives. To avoid the need for higher-order derivatives, the Runge-Kutta method evaluates the function $F(x,y)$ at more points.

Again, consider the first-order differential equation (6.1):

$$y' = F(x, y); \qquad y(x_0) = y_0$$

Algorithm (6.7) is rewritten in the following way:

$$y_{i+1} = y_i + \frac{1}{2} \Delta x \, F(x_i, y_i) + \frac{1}{2} \Delta x \left\{ F + \Delta x \frac{\partial F}{\partial x} + \Delta x \frac{\partial F}{\partial y} F \right\}_{(x_i,\, y_i)}$$

According to Taylor's formula, we have, for the terms in Δx

$$\left\{ F + \Delta x \frac{\partial F}{\partial x} + \Delta x \frac{\partial F}{\partial y} F \right\}_{(x_i,\, y_i)} = F(x_i + \Delta x, y_i + \Delta x \, F(x_i, y_i))$$

Thus, we obtain the ***second-order Runge-Kutta*** algorithm:

$$y_{i+1} = y_i + \frac{1}{2} \Delta x \, (M_1 + M_2), \qquad i = 0, 1, \ldots, n-1 \tag{6.8}$$

where

$$M_1 = F(x_i, y_i)$$

$$M_2 = F(x_i + \Delta x, y_i + \Delta x \, M_1)$$

The most commonly used ***Runge-Kutta*** method is that of ***fourth order***, consisting of the following algorithm:

$$y_{i+1} = y_i + \frac{1}{6} \Delta x \, (N_1 + 2N_2 + 2N_3 + N_4), \qquad i = 0, 1, \ldots, n-1 \tag{6.9}$$

where

$$N_1 = F(x_i, y_i)$$

$$N_2 = F\left(x_i + \frac{1}{2}\Delta x,\ y_i + \frac{1}{2}\Delta x N_1\right)$$

$$N_3 = F\left(x_i + \frac{1}{2}\Delta x,\ y_i + \frac{1}{2}\Delta x N_2\right)$$

$$N_4 = F(x_i + \Delta x,\ y_i + \Delta x N_3)$$

It is easily verified that formula (6.9) coincides with Taylor's formula for $y(x_i + \Delta x)$ for terms up to Δx^4.

6.3.6 Example: Solution of an equation using the Runge-Kutta method

Consider the equation:

$$y' = y - \frac{2x}{y}; \qquad y(0) = 1 \tag{a}$$

We wish to approximate the solution of equation (a) in $x_1 = 0.2$.

According to the second-order method (6.8), we have

$$y_1 = y_0 + \frac{1}{2}\Delta x\,(M_1 + M_2)$$

with $x_0 = 0$, $y_0 = 1$, $\Delta x = 0.2$. Because

$$M_1 = F(0; 1) = 1$$

$$M_2 = F(0.2; 1.2) = 0.866\,667$$

we find

$$y(0.2) \approx y_1 = 1.186\,667$$

According to the fourth-order Runge-Kutta method (6.9), we have

$$y_1 = y_0 + \frac{1}{6}\Delta x\,(N_1 + 2N_2 + 2N_3 + N_4)$$

with $x_0 = 0$, $y_0 = 1$, $\Delta x = 0.2$. Because

$N_1 = F(0; 1) = 1$

$N_2 = F(0.1; 1.1) = 0.918182$

$N_3 = F(0.1; 1.091\,818) = 0.908637$

$N_4 = F(0.2; 1.181\,727) = 0.843239$

we find

$$y(0.2) \approx y_1 = 1.183229$$

The exact solution of equation (a) is given by $y = \sqrt{2x+1}$, which gives $y(0.2) = 1.183216$.

6.4 NUMERICAL INTEGRATION METHODS FOR A SYSTEM OF TWO FIRST-ORDER EQUATIONS

6.4.1 Method using the Taylor method

We wish to approximate the functions $x(t)$ and $y(t)$ satisfying the system of first-order differential equations:

$$\begin{aligned} \dot{x} &= F(x, y, t) \\ \dot{y} &= G(x, y, t) \end{aligned} \tag{6.10}$$

subject to the initial conditions $x(t_0) = x_0$, $y(t_0) = y_0$.

Using the Taylor-series expansions:

$$x_{i+1} = x_i + \Delta t\, \dot{x}_i + \frac{1}{2!} \Delta t^2\, \ddot{x}_i + \ldots$$

$$y_{i+1} = y_i + \Delta t\, \dot{y}_i + \frac{1}{2!} \Delta t^2\, \ddot{y}_i + \ldots$$

and the relations:

$$\ddot{x} = \frac{d}{dt}\{F(x, y, t)\} = \frac{\partial F}{\partial x}\dot{x} + \frac{\partial F}{\partial y}\dot{y} + \frac{\partial F}{\partial t}$$

$$\ddot{y} = \frac{d}{dt}\{G(x, y, t)\} = \frac{\partial G}{\partial x}\dot{x} + \frac{\partial G}{\partial y}\dot{y} + \frac{\partial G}{\partial t}$$

the solution of (6.10) can be approximated by the following algorithm:

$$x_{i+1} = x_i + \Delta t\, F(x_i, y_i, t_i) + \frac{1}{2}\Delta t^2 \left(\frac{\partial F}{\partial x} F + \frac{\partial F}{\partial y} G + \frac{\partial F}{\partial t} \right)_{(x_i,\, y_i,\, t_i)}$$

$$y_{i+1} = y_i + \Delta t\, G(x_i, y_i, t_i) + \frac{1}{2}\Delta t^2 \left(\frac{\partial G}{\partial x} F + \frac{\partial G}{\partial y} G + \frac{\partial G}{\partial t} \right)_{(x_i,\, y_i,\, t_i)}$$

$$i = 0, 1, \ldots, n-1 \qquad \textbf{(6.11)}$$

6.4.2 Example: Solution of a system of equations using the Taylor method

We again consider equation (a) from section 6.2.2 (movement of the pendulum):

$$a\, \ddot{x} + g \sin x = 0$$

If we set

$$\dot{x} = y = F(x, y)$$

$$\dot{y} = \ddot{x} = -\frac{g}{a} \sin x = G(x, y)$$

we obtain from formula (6.11):

$$x_{i+1} = x_i + \Delta t\, y_i - \frac{1}{2}\Delta t^2 \frac{g}{a} \sin x_i$$

$$y_{i+1} = y_i - \Delta t \frac{g}{a} \sin x_i - \frac{1}{2}\Delta t^2 \frac{g}{a} y_i \cos x_i$$

If we assumed that the initial conditions are $x(0) = \pi/4, \dot{x}(0) = 0$, the algorithm starts from $x_0 = \pi/4, y_0 = 0$.

6.4.3 Runge-Kutta Method

Consider the system of differential equations:

$$\left.\begin{aligned} \dot{x} &= F(x, y, t) \\ \dot{y} &= G(x, y, t); \\ x(t_0) &= x_0,\ y(t_0) = y_0 \end{aligned}\right\} \qquad (6.12)$$

According to the fourth-order Runge-Kutta method, the solution of (6.12) is approximated by using the algorithm:

$$\left.\begin{aligned} x_{i+1} &= x_i + \frac{1}{6}\Delta t\,(H_1 + 2H_2 + 2H_3 + H_4) \\ y_{i+1} &= y_i + \frac{1}{6}\Delta t\,(L_1 + 2L_2 + 2L_3 + L_4) \\ i &= 0, 1, \ldots, n-1 \end{aligned}\right\} \qquad \textbf{(6.13)}$$

where

$$H_1 = F(x_i, y_i, t_i)$$
$$L_1 = G(x_i, y_i, t_i)$$
$$H_2 = F\left(x_i + \frac{1}{2}\Delta t\, H_1,\ y_i + \frac{1}{2}\Delta t\, L_1,\ t_i + \frac{1}{2}\Delta t\right)$$
$$L_2 = G\left(x_i + \frac{1}{2}\Delta t\, H_1,\ y_i + \frac{1}{2}\Delta t\, L_1,\ t_i + \frac{1}{2}\Delta t\right)$$
$$H_3 = F\left(x_i + \frac{1}{2}\Delta t\, H_2,\ y_i + \frac{1}{2}\Delta t\, L_2,\ t_i + \frac{1}{2}\Delta t\right)$$
$$L_3 = G\left(x_i + \frac{1}{2}\Delta t\, H_2,\ y_i + \frac{1}{2}\Delta t\, L_2,\ t_i + \frac{1}{2}\Delta t\right)$$
$$H_4 = F(x_i + \Delta t\, H_3,\ y_i + \Delta t\, L_3,\ t_i + \Delta t)$$
$$L_4 = G(x_i + \Delta t\, H_3,\ y_i + \Delta t\, L_3,\ t_i + \Delta t)$$

6.4.4 Example: Solution of a system of two differential equations by the fourth-order Runge-Kutta method

Consider the system:

$$\dot{x} = xy + t$$
$$\dot{y} = ty + x$$
$$x(0) = 1,\ y(0) = -1$$

We wish to approximate the solution of this system at $t_1 = 0.2$.
Because

$$F(x, y, t) = xy + t$$
$$G(x, y, t) = ty + x$$
$$\Delta t = 0.2$$

we calculate

$$H_1 = -1$$
$$L_1 = 1$$
$$H_2 = 0.1 + 0.9\,(-0.9) = -0.71$$
$$L_2 = 0.1\,(-1 + 0.1) + (1 - 0.1) = 0.81$$
$$H_3 = (1 - 0.071)\,(-1 + 0.081) + 0.1 = -0.754$$
$$L_3 = 0.1\,(-1 + 0.081) + (1 - 0.071) = 0.837$$
$$H_4 = (1 - 0.1508)\,(-1 + 0.1674) + 0.2 = -0.507$$
$$L_4 = 0.2\,(-1 + 0.1674) + (1 - 0.1508) = 0.683$$

and from formula (6.13) we find

$$x\,(0.2) \approx 0.8522$$
$$y\,(0.2) \approx -0.8341$$

6.4.5 Example: Ballistics problem

We wish to study the ballistic trajectory of a shell P fired with the initial velocity v_0 at the angle φ_0 with respect to the x-axis (see fig. 6.9).

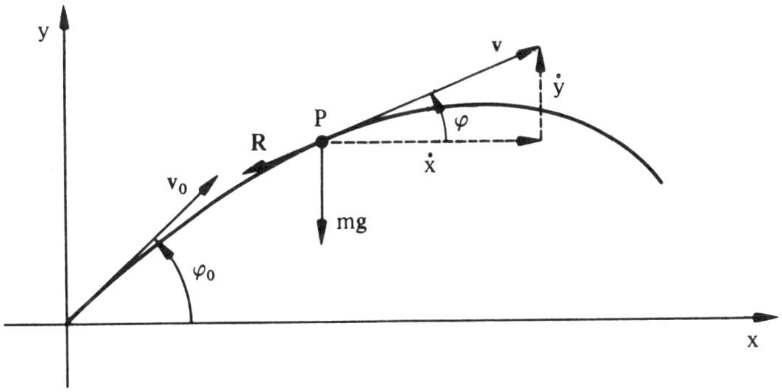

Fig. 6.9

If we take the air resistance R into account, the movement of P is described by the differential equation:

$$m\ddot{\mathbf{x}} = \mathbf{R} + m\mathbf{g} \qquad \text{(a)}$$

Assuming that R is proportional to the velocity v squared, i.e., $R = \alpha v^2$, and writing the vector equation (a) for each of its components, we obtain the differential system:

$$m\ddot{x} = -\alpha v^2 \cos\varphi = -\alpha\sqrt{\dot{x}^2+\dot{y}^2}\,\dot{x}$$
$$m\ddot{y} = -\alpha v^2 \sin\varphi - mg = -\alpha\sqrt{\dot{x}^2+\dot{y}^2}\,\dot{y} - mg \tag{b}$$

When we include the numerical data $m = 1$ kg, $g = 10$ m/sec^2, $\alpha = 10^{-3}$ kg/m, system (b) becomes

$$\ddot{x} = -10^{-3}\sqrt{\dot{x}^2+\dot{y}^2}\,\dot{x}$$
$$\ddot{y} = -10^{-3}\sqrt{\dot{x}^2+\dot{y}^2}\,\dot{y} - 10 \tag{c}$$

If we set $\dot{x} = u$, $\dot{y} = w$. System (c) is thus equivalent to the system:

$$\dot{u} = -10^{-3}\sqrt{u^2+w^2}\,u = F(u, w)$$
$$\dot{w} = -10^{-3}\sqrt{u^2+w^2}\,w - 10 = G(u, w) \tag{d}$$

Assuming that, for $t_0 = 0$, $v_0 = 1000$ m/sec and $\varphi_0 = 20°$ is the elevation angle of firing, we have

$$u_0 = 1000 \cos 20° = 939.69$$
$$w_0 = 1000 \sin 20° = 342.02$$

For $\Delta t = 0.5$ sec, by applying the fourth-order Runge-Kutta algorithm (6.13), we calculate

$$\dot{x}(0.05) \approx u_1 = 939.69 + \frac{0.05}{6}(-939.69 - 2\cdot 893.22 - 2\cdot 895.48 - 852.13)$$
$$= 894.95$$

$$\dot{y}(0.05) \approx w_1 = 342.02 + \frac{0.05}{6}(-352.02 - 2\cdot 334.86 - 2\cdot 335.69 - 319.69)$$
$$= 325.25$$

$$\dot{x}(0.1) \approx u_2 = 894.95 + \frac{0.05}{6}(-852.19 - 2\cdot 812.03 - 2\cdot 813.9 - 776.32)$$
$$= 854.28$$

$$\dot{y}(0.1) \approx w_2 = 325.25 + \frac{0.05}{6}(-319.71 - 2\cdot 304.88 - 2\cdot 305.57 - 291.69)$$
$$= 309.98$$

The position of the shell after a tenth of a second, according to Simpson's formula (5.12), is

$$x(0.1) \approx \frac{0.05}{3}(u_0 + 4u_1 + u_2) = 89.56 \text{ m}$$

$$y(0.1) \approx \frac{0.05}{3}(w_0 + 4w_1 + w_2) = 32.55 \text{ m}$$

6.5 EXERCISES

6.5.1 Consider the first-order differential equation:

$$y' = -x/y$$

- integrate the equation analytically and sketch the trajectories in the xy-plane for the three initial conditions $y(0) = 0$, $y(0) = 1$, $y(0) = 2$;
- determine the four isoclines of slope $y' = 0$, $y' = \infty$, $y' = 1$, $y' = -1$ and sketch the trajectories for the same initial conditions.

6.5.2 Given the second-order autonomous differential equation:

$$\ddot{x} + \dot{x}^2 + x\dot{x} = 0$$

- reduce the equation to a first-order equation by setting $\dot{x} = y(x)$. Integrate the resulting equation analytically and sketch the trajectory in the phase plane $x, \dot{x}$ for the initial conditions $x(0) = 0$, $\dot{x}(0) = 2$;
- determine the isoclines with slopes $y' = 0$, $y' = -1$, $y' = -3/2$, $y' = -2$ and sketch the trajectory subject to the same initial conditions.

6.5.3 Solve equation (a) of section 6.3.2 with the aid of Taylor's method (6.7) and compare the results.

6.5.4 Consider a sphere of mass m in a liquid (see fig. 6.10),

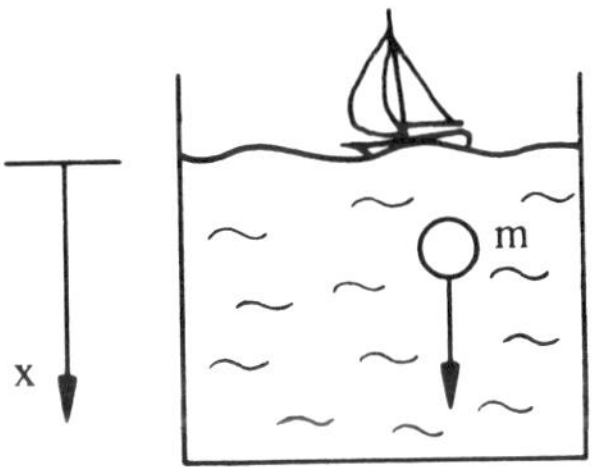

Fig. 6.10

- set up the differential equation of motion of the sphere, assuming that the resistance R of the liquid is proportional to its speed squared, i.e., $R = \alpha \dot{x}^2$; m = 1 kg, g = 10 m/sec², $\alpha = 1$ kg/m;
- set $\dot{x} = y(x)$ and integrate the resulting equation subject to the initial conditions $x(0) = \dot{x}(0) = 0$. Sketch the trajectory in the phase plane x, $\dot{x}$. Determine the final speed ($t \to \infty$) of the sphere;
- determine x(t) analytically by integrating the equation thus obtained (substitute $x = \ln u$);
- transform the differential equation of motion of the sphere into a first-order differential system. Give the algorithm using Taylor's formula for numerical integration of this system. For $\Delta t = 0.01$ sec, approximate x(0.02) and $\dot{x}(0.02)$.

6.5.5 Given the differential system:

$$\dot{x} = xy + t^2$$

$$\dot{y} = x + y + t$$

$$x(1) = 2, \; y(1) = 2$$

- give the algorithm using Taylor's formula for numerical integration of this system;
- for $\Delta t = 0.02$ sec, approximate x(1.04) and y(1.04).

Part II
Vector Analysis

Chapter 7

Vector Differentiation and Differential Operators

7.1 REVIEW OF VECTOR ALGEBRA

In a rectangular xyz coordinate system, any vector **F** can be placed in the form:

$$\mathbf{F} = F_x\,\mathbf{i} + F_y\,\mathbf{j} + F_z\,\mathbf{k} = (F_x, F_y, F_z), \tag{7.1}$$

where **i, j, k** designate the rectangular unit vectors and F_x, F_y, F_z are the components of the vector **F** in this coordinate system. If the vector **F** has its initial point at the origin of the xyz coordinate system, the components F_x, F_y, and F_z are the orthogonal projections of **F** onto the coordinate axes and give the coordinates of the end point P of the vector **F** (fig. 7.1).

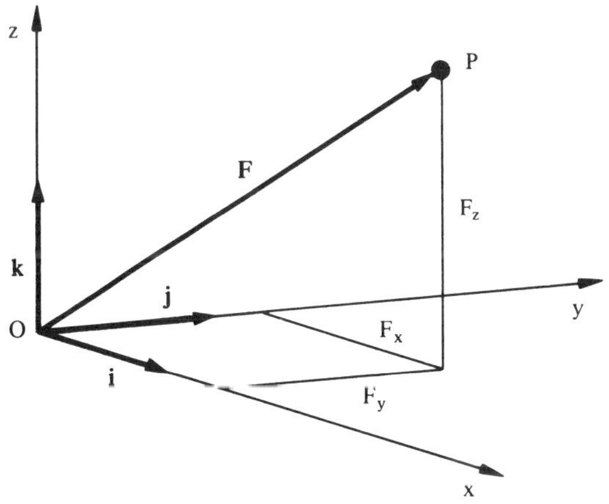

Fig. 7.1

The ***magnitude*** of the vector **F** is written $|\mathbf{F}|$. Knowing the components F_x, F_y, F_z of **F**, we have for its value:

$$|\mathbf{F}| = \sqrt{F_x^2 + F_y^2 + F_z^2}$$

Consider two vectors **F** and **G**. By placing the initial point of **G** on the end point of **F** and joining the initial point of **F** to the end point of **G**, we obtain the ***vector sum*** of the two vectors **F** and **G** (see fig. 7.2), which is written **S** = **F** + **G**.

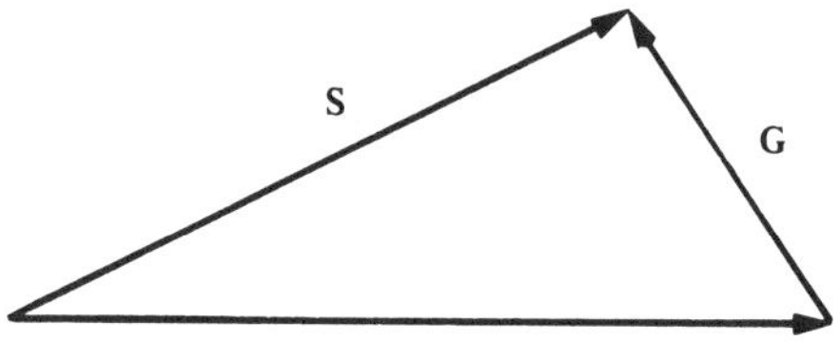

Fig. 7.2

The vector sum obeys the commutative law and the associative law for addition, expressed respectively by the formulas:

$$\mathbf{F} + \mathbf{G} = \mathbf{G} + \mathbf{F} \tag{7.2}$$

$$\mathbf{F} + (\mathbf{G} + \mathbf{H}) = (\mathbf{F} + \mathbf{G}) + \mathbf{H} \tag{7.3}$$

where **H** is a third vector. Knowing the components of two vectors **F** and **G**, we have for the vector sum **S**:

$$\mathbf{S} = (F_x + G_x)\,\mathbf{i} + (F_y + G_y)\,\mathbf{j} + (F_z + G_z)\,\mathbf{k} \tag{7.4}$$

Let us now consider a vector **F** and a scalar ϕ. The ***product of the vector*** **F** ***by the scalar*** ϕ is a vector having the same direction as **F**, but whose magnitude is $|\phi|$ multiplied by $|\mathbf{F}|$. The representation of $\phi\,\mathbf{F}$ by means of the components of **F** is given by

$$\phi\mathbf{F} = \phi F_x\,\mathbf{i} + \phi F_y\,\mathbf{j} + \phi F_z\,\mathbf{k} \tag{7.5}$$

The ***scalar product*** **F·G** of two vectors **F** and **G** is a ***scalar*** equal to the product of the magnitudes of these two vectors multiplied by the cosine of the angle (**F**, **G**) included between them:

$$\mathbf{F} \cdot \mathbf{G} = |\mathbf{F}|\,|\mathbf{G}| \cos(\mathbf{F}, \mathbf{G}) \tag{7.6}$$

The scalar product is expressed in terms of the components of the two vectors as follows:

$$\mathbf{F} \cdot \mathbf{G} = F_x G_x + F_y G_y + F_z G_z \tag{7.7}$$

i.e., the scalar product is equal to the sum of the products of the corresponding components of **F** and **G**. We can easily see that the scalar product is distributive with respect to the vector sum:

$$(\mathbf{F} + \mathbf{G}) \cdot \mathbf{H} = \mathbf{F} \cdot \mathbf{H} + \mathbf{G} \cdot \mathbf{H} \tag{7.8}$$

The ***vector product*** $\mathbf{F} \wedge \mathbf{G}$ of two vectors **F** and **G** is a vector whose magnitude is equal to the surface of the parallelogram constructed by these two vectors, and whose direction is perpendicular to the plane of this parallelogram such that the three vectors form a right-handed system. Taking this definition into account, the magnitude of the vector product equals

$$|\mathbf{F} \wedge \mathbf{G}| = |\mathbf{F}|\,|\mathbf{G}| \sin(\mathbf{F}, \mathbf{G}) \tag{7.9}$$

Let us note that the vector product is not commutative:

$$\mathbf{F} \wedge \mathbf{G} = -\mathbf{G} \wedge \mathbf{F}$$

The vector product is expressed in terms of the components of the two vectors as follows:

$$\mathbf{F} \wedge \mathbf{G} = (F_y G_z - F_z G_y)\mathbf{i} + (F_z G_x - F_x G_z)\mathbf{j} + (F_x G_y - F_y G_x)\mathbf{k} \tag{7.10}$$

or written in convenient determinant form:

$$\mathbf{F} \wedge \mathbf{G} = \begin{vmatrix} \mathbf{i} & \mathbf{j} & \mathbf{k} \\ F_x & F_y & F_z \\ G_x & G_y & G_z \end{vmatrix} \tag{7.11}$$

Using relation (7.10), we can easily verify that the vector product is distributive with respect to the vector sum:

$$(\mathbf{F} + \mathbf{G}) \wedge \mathbf{H} = \mathbf{F} \wedge \mathbf{H} + \mathbf{G} \wedge \mathbf{H} \tag{7.12}$$

When the ***volume of the parallelepiped*** constructed by the three vectors **F**,**G**, and **H** is calculated, the following relations for the ***triple mixed product*** are obtained:

$$\mathbf{F} \cdot (\mathbf{G} \wedge \mathbf{H}) = \mathbf{G} \cdot (\mathbf{H} \wedge \mathbf{F}) = \mathbf{H} \cdot (\mathbf{F} \wedge \mathbf{G}) \tag{7.13}$$

Using the components of vectors **F**, **G**, and **H**, the volume V of the parallelepiped formed by these vectors is obtained as follows:

$$V = \mathbf{F} \cdot (\mathbf{G} \wedge \mathbf{H}) = \begin{vmatrix} F_x & F_y & F_z \\ G_x & G_y & G_z \\ H_x & H_y & H_z \end{vmatrix} \tag{7.14}$$

Finally, we have for the ***triple vector product*** $\mathbf{F} \wedge (\mathbf{G} \wedge \mathbf{H})$ the formula:

$$\mathbf{F} \wedge (\mathbf{G} \wedge \mathbf{H}) = (\mathbf{F} \cdot \mathbf{H})\,\mathbf{G} - (\mathbf{F} \cdot \mathbf{G})\,\mathbf{H} \tag{7.15}$$

7.2 DIFFERENTIATION OF VECTORS

Let the components of the vector **F** = **F** (u) be functions of the real variable u. If the vector **F** has its initial point at the origin of the xyz coordinate system, and if the parameter u is varied, the vector **F** describes a certain curve in space (fig. 7.3).

Consider the terminal points P_0 and P_1 of the vector **F** for the parameter values u_0 and $u_0 + \Delta u$. The segment P_0P_1 corresponds to the vector difference $\Delta\mathbf{F} = \mathbf{F}(u_0 + \Delta u) - \mathbf{F}(u_0)$, and the quotient:

$$\frac{\Delta \mathbf{F}}{\Delta u} = \frac{\mathbf{F}(u_0 + \Delta u) - \mathbf{F}(u_0)}{\Delta u}$$

defines a vector parallel to the segment $\overline{P_0P_1}$. The limit of this vector for $\Delta u \to 0$, if it exists, is called the ***derivative*** of **F** (u) at $u = u_0$:

$$\frac{d\mathbf{F}}{du}(u_0) = \lim_{\Delta u \to 0} \frac{\mathbf{F}(u_0 + \Delta u) - \mathbf{F}(u_0)}{\Delta u} \tag{7.16}$$

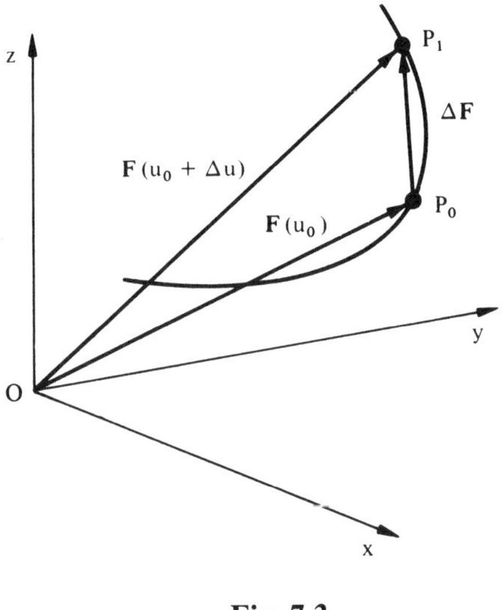

Fig. 7.3

The derivative is clearly a vector parallel to the tangent to the curve at P_0. If the components of the vector function **F** are known:

$$\mathbf{F}(u) = F_x(u)\,\mathbf{i} + F_y(u)\,\mathbf{j} + F_z(u)\,\mathbf{k}$$

the definition of the derivative (7.16) gives:

$$\frac{d\mathbf{F}}{du} = \frac{dF_x}{du}\mathbf{i} + \frac{dF_y}{du}\mathbf{j} + \frac{dF_z}{du}\mathbf{k} \tag{7.17}$$

i.e., ***the derivative of a vector is obtained by differentiation of its components.***

The usual rules of differentiation are generalized to the differentiation of the sum of two vectors, of a vector multiplied by a scalar, and the scalar and vector products of two vectors. We have the following formulas:

$$\frac{d}{du}(\mathbf{F} + \mathbf{G}) = \frac{d\mathbf{F}}{du} + \frac{d\mathbf{G}}{du} \tag{7.18}$$

$$\frac{d}{du}(\phi\,\mathbf{F}) = \frac{d\phi}{du}\mathbf{F} + \phi\frac{d\mathbf{F}}{du} \tag{7.19}$$

$$\frac{d}{du}(\mathbf{F} \cdot \mathbf{G}) = \frac{d\mathbf{F}}{du} \cdot \mathbf{G} + \mathbf{F} \cdot \frac{d\mathbf{G}}{du} \tag{7.20}$$

$$\frac{d}{du}(\mathbf{F} \wedge \mathbf{G}) = \frac{d\mathbf{F}}{du} \wedge \mathbf{G} + \mathbf{F} \wedge \frac{d\mathbf{G}}{du} \tag{7.21}$$

7.3 GRADIENT, DIVERGENCE, AND CURL

If a physical quantity (for example, temperature, pressure, *et cetera*) takes a scalar value at each point (x,y,z) of a coordinate system, it is said to define a ***scalar field***, designated by ϕ (x,y,z).

If the physical quantity being considered (for example, an acceleration, a force, *et cetera*) corresponds to a vector at each point (x,y,z) of a coordinate system, this vector defines a ***vector field***, denoted by **F** (x,y,z).

We will first consider a scalar field ϕ(x,y,z) with continuous first partial derivatives. Then, the ***gradient*** of ϕ, written ***grad*** ϕ, is defined by

$$\mathbf{grad}\,\phi = \frac{\partial\phi}{\partial x}\mathbf{i} + \frac{\partial\phi}{\partial y}\mathbf{j} + \frac{\partial\phi}{\partial z}\mathbf{k} \tag{7.22}$$

Let us note that ***grad*** ϕ defines a ***vector field.***

Now consider the vector field **F** (x,y,z) whose components again have continuous first partial derivatives. Then, the ***divergence*** of **F**, written div **F**, is defined by

$$\text{div}\,\mathbf{F} = \frac{\partial F_x}{\partial x} + \frac{\partial F_y}{\partial y} + \frac{\partial F_z}{\partial z} \tag{7.23}$$

where F_x, F_y, and F_z are the components of **F**. Let us note that div **F** defines a ***scalar field***.

The *curl* of **F**, written *curl* **F**, is defined by

$$\mathbf{curl\ F} = \left(\frac{\partial F_z}{\partial y} - \frac{\partial F_y}{\partial z}\right)\mathbf{i} + \left(\frac{\partial F_x}{\partial z} - \frac{\partial F_z}{\partial x}\right)\mathbf{j} + \left(\frac{\partial F_y}{\partial x} - \frac{\partial F_x}{\partial y}\right)\mathbf{k} \quad (7.24)$$

or, using a determinant:

$$\mathbf{curl\ F} = \begin{vmatrix} \mathbf{i} & \mathbf{j} & \mathbf{k} \\ \dfrac{\partial}{\partial x} & \dfrac{\partial}{\partial y} & \dfrac{\partial}{\partial z} \\ F_x & F_y & F_z \end{vmatrix} \quad (7.25)$$

Let us note that **curl F** defines a ***vector field***.

Introducing the ***vector operator*** ∇, defined by

$$\nabla = \mathbf{i}\frac{\partial}{\partial x} + \mathbf{j}\frac{\partial}{\partial y} + \mathbf{k}\frac{\partial}{\partial z} \quad (7.26)$$

definitions (7.22), (7.23), and (7.24) can be written in the following way:

$$\mathbf{grad}\ \phi = \nabla\phi \quad (7.27)$$

$$\text{div}\ \mathbf{F} = \nabla \cdot \mathbf{F} \quad (7.28)$$

$$\mathbf{curl\ F} = \nabla \wedge \mathbf{F} \quad (7.29)$$

i.e., the gradient of ϕ is the product of the operator ∇ and the scalar ϕ, the divergence of **F** is the scalar product of ∇ and the vector **F** and the curl of **F** is the vector product of ∇ and the vector **F**. It is clear that the use of the operator ∇ permits the divergence, gradient, and curl to be expressed very succinctly.

7.4 SOME DIFFERENTIATION FORMULAS

A vector field **F** (x,y,z) is called ***conservative***, if the vector **F** is the gradient of a certain scalar function ϕ (x,y,z):

$$\mathbf{F} = \mathbf{grad}\ \phi \quad (7.30)$$

In this case, we say that **F** derives from the potential ϕ. It is easy to verify that ***the curl of a conservative vector field vanishes***. In effect, we have

$$\textbf{curl grad}\ \phi = \begin{vmatrix} \mathbf{i} & \mathbf{j} & \mathbf{k} \\ \dfrac{\partial}{\partial x} & \dfrac{\partial}{\partial y} & \dfrac{\partial}{\partial z} \\ \dfrac{\partial \phi}{\partial x} & \dfrac{\partial \phi}{\partial y} & \dfrac{\partial \phi}{\partial z} \end{vmatrix}$$

$$= \mathbf{i}\left(\frac{\partial^2 \phi}{\partial y \partial z} - \frac{\partial^2 \phi}{\partial z \partial y}\right) + \mathbf{j}\left(\frac{\partial^2 \phi}{\partial z \partial x} - \frac{\partial^2 \phi}{\partial x \partial z}\right) + \mathbf{k}\left(\frac{\partial^2 \phi}{\partial x \partial y} - \frac{\partial^2 \phi}{\partial y \partial x}\right) = \mathbf{0}$$

with the hypothesis that the second partial derivatives are continuous so that the order of differentiation can be interchanged.

The ***differential operator***:

$$\Delta = \frac{\partial^2}{\partial x^2} + \frac{\partial^2}{\partial y^2} + \frac{\partial^2}{\partial z^2} \tag{7.31}$$

is called the ***Laplace operator***. Applying the Laplace operator Δ to the scalar field ϕ, we have the relationship:

$$\Delta \phi = \text{div}\ \textbf{grad}\ \phi = \nabla \cdot \nabla \phi \tag{7.32}$$

It is easily verified that ***the divergence of the curl of a vector field cancels out:***

$$\text{div}\ \textbf{curl}\ \mathbf{F} = \nabla \cdot (\nabla \wedge \mathbf{F})$$

$$= \frac{\partial}{\partial x}\left(\frac{\partial F_z}{\partial y} - \frac{\partial F_y}{\partial z}\right) + \frac{\partial}{\partial y}\left(\frac{\partial F_x}{\partial z} - \frac{\partial F_z}{\partial x}\right) + \frac{\partial}{\partial z}\left(\frac{\partial F_y}{\partial x} - \frac{\partial F_x}{\partial y}\right)$$

$$= \frac{\partial^2 F_x}{\partial y \partial z} - \frac{\partial^2 F_x}{\partial z \partial y} + \frac{\partial^2 F_y}{\partial z \partial x} - \frac{\partial^2 F_y}{\partial x \partial z} + \frac{\partial^2 F_z}{\partial x \partial y} - \frac{\partial^2 F_z}{\partial y \partial x} = 0$$

The following identities are of frequent use and they can be verified by direct expansion:

$$\text{div}\ (\phi\ \mathbf{F}) - \phi\ \text{div}\ \mathbf{F} + \mathbf{F} \cdot \textbf{grad}\ \phi \tag{7.33}$$

$$\text{div}\ (\mathbf{F} \wedge \mathbf{G}) = \mathbf{G} \cdot \textbf{curl}\ \mathbf{F} - \mathbf{F} \cdot \textbf{curl}\ \mathbf{G} \tag{7.34}$$

$$\textbf{curl}\ (\phi\ \mathbf{F}) = \textbf{grad}\ \phi \wedge \mathbf{F} + \phi\ \textbf{curl}\ \mathbf{F} \tag{7.35}$$

$$\textbf{curl curl}\ \mathbf{F} = \textbf{grad}\ \text{div}\ \mathbf{F} - \Delta\ \mathbf{F} \tag{7.36}$$

In formula (7.36), the Laplace operator Δ is applied to the vector field **F**; $\Delta\mathbf{F}$ designates the vector with components ΔF_x, ΔF_y, ΔF_z. We verify the last formula (7.36) and leave the task of proving the others to the reader. We can see that each component on the left side of (7.36) coincides with the corresponding component on the right side. As an example, we shall consider the x component:

$$(\mathbf{curl\ curl\ F})_x = \frac{\partial}{\partial y}\left(\frac{\partial F_y}{\partial x} - \frac{\partial F_x}{\partial y}\right) - \frac{\partial}{\partial z}\left(\frac{\partial F_x}{\partial z} - \frac{\partial F_z}{\partial x}\right)$$

Then, adding and subtracting the term $\partial^2 F_x/\partial x^2$ on the right-hand side of the last relation, we obtain

$$(\mathbf{curl\ curl\ F})_x = \frac{\partial}{\partial x}\left(\frac{\partial F_x}{\partial x} + \frac{\partial F_y}{\partial y} + \frac{\partial F_z}{\partial z}\right) - \left(\frac{\partial^2 F_x}{\partial x^2} + \frac{\partial^2 F_x}{\partial y^2} + \frac{\partial^2 F_x}{\partial z^2}\right)$$

$$= \frac{\partial}{\partial x}(\text{div}\ \mathbf{F}) - \Delta F_x$$

Thus, we find

$$\mathbf{curl\ curl\ F} = \mathbf{grad}\ \text{div}\ \mathbf{F} - \Delta\mathbf{F}$$

7.5 TECHNICAL APPLICATIONS

7.5.1 Example: Force due to pressure in a fluid

In a perfect fluid, the forces due to pressure are normal to the surface (no internal friction) of an element of fluid in the form of a rectangular parallelepiped with edges parallel to the coordinate axes. The pressure is described by the scalar field p(x,y,z), which is the force exerted on a unit surface at the point (x,y,z). The differential dp_z over a differential height dz will be

$$dp_z = \frac{\partial p}{\partial z}\,dz$$

from which we get the differential force dF_z exerted by the pressure distribution on the two faces parallel to the xy-plane with differential surfaces dxdy (fig. 7.4):

$$d\mathbf{F}_z = -\mathbf{k}\frac{\partial p}{\partial z}\,dxdydz$$

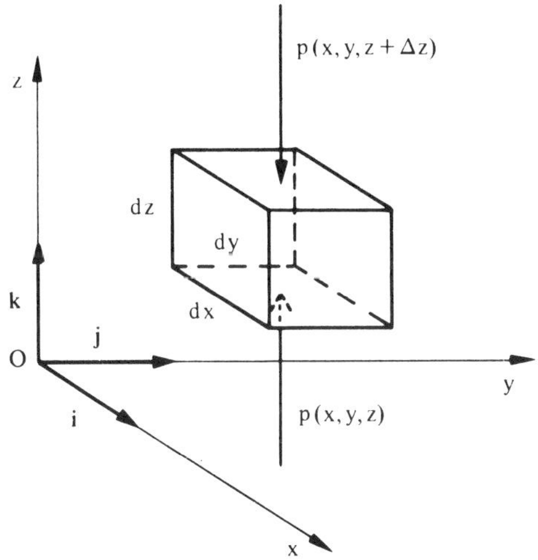

Fig. 7.4

Repeating this procedure for the two other axes, we find for the resultant **dF** of all the pressure forces acting on the surface of the elementary volume dV:

$$d\mathbf{F} = -\mathbf{grad}\, p\, dV \tag{7.37}$$

7.5.2 Example: Derivative of a scalar field in a given direction

Consider a scalar field ϕ (x,y,z) and the unit vector $\mathbf{n} = (\cos\alpha, \cos\beta, \cos\gamma)$. Suppose that the vector **n** has its initial point at the point (x,y,z), and consider the point (x + Δx, y + Δy, z + Δz) on **n** at the distance Δs from the point (x,y,z). Therefore, we have

$$\Delta x^2 + \Delta y^2 + \Delta z^2 = \Delta s^2$$

The total change of the function ϕ along the differential displacement Δs on **n** will be

$$\Delta\phi = \frac{\partial\phi}{\partial x}\Delta x + \frac{\partial\phi}{\partial y}\Delta y + \frac{\partial\phi}{\partial z}\Delta z + \epsilon_x \Delta x + \epsilon_y \Delta y + \epsilon_z \Delta z \tag{7.38}$$

where the terms ϵ_x, ϵ_y, and ϵ_z converge toward zero as $\Delta s \to 0$. Dividing relation (7.38) by Δs:

$$\frac{\Delta\phi}{\Delta s} = \frac{\partial\phi}{\partial x}\frac{\Delta x}{\Delta s} + \frac{\partial\phi}{\partial y}\frac{\Delta y}{\Delta s} + \frac{\partial\phi}{\partial z}\frac{\Delta z}{\Delta s} + \epsilon_x\frac{\Delta x}{\Delta s} + \epsilon_y\frac{\Delta y}{\Delta s} + \epsilon_z\frac{\Delta z}{\Delta s}$$

and taking into account the relations:

$$\frac{\Delta x}{\Delta s} = \cos\alpha, \qquad \frac{\Delta y}{\Delta s} = \cos\beta, \qquad \frac{\Delta z}{\Delta s} = \cos\gamma$$

we obtain, on reaching the limit as $\Delta s \to 0$:

$$\lim_{\Delta s \to 0} \frac{\Delta\phi}{\Delta s} = \frac{\partial\phi}{\partial x}\cos\alpha + \frac{\partial\phi}{\partial y}\cos\beta + \frac{\partial\phi}{\partial z}\cos\gamma = \mathbf{grad}\ \phi \cdot \mathbf{n} \tag{7.39}$$

This limit is termed the ***derivative of the scalar function*** ϕ (x,y,z) ***at the point*** (x,y,z) ***in the direction of the unit vector*** **n** and is written $\partial\phi/\partial n$.

7.5.3 Example: Deformation of a nonrigid body

Let us consider a homogeneous linear deformation of a rectangular parallelepiped for which the components of the displacement vector **F** are homogeneous linear functions of the coordinates x,y,z:

$$F_x = \epsilon_{xx}\, x + \epsilon_{xy}\, y + \epsilon_{xz}\, z$$

$$F_y = \epsilon_{yx}\, x + \epsilon_{yy}\, y + \epsilon_{yz}\, z$$

$$F_z = \epsilon_{zx}\, x + \epsilon_{zy}\, y + \epsilon_{zz}\, z$$

Each point of the parallelepiped is displaced by the vector **F**. After deformation, the point (x,y,z) will have the coordinates given by:

$$\xi = (1 + \epsilon_{xx})\, x + \epsilon_{xy}\, y + \epsilon_{xz}\, z$$

$$\eta = \epsilon_{yx}\, x + (1 + \epsilon_{yy})\, y + \epsilon_{yz}\, z$$

$$\zeta = \epsilon_{zx}\, x + \epsilon_{zy}\, y + (1 + \epsilon_{zz})\, z$$

We shall now determine the change of the volume of this body produced by this deformation, assuming that the coefficients ϵ_{xx}, $\epsilon_{xy}, \ldots$, are small. It is sufficient to study the deformation of the cube constructed by the three unit vectors **i,j,k**, which are transformed into a parallelepiped with the edges:

$$\mathbf{i}^* = (1 + \epsilon_{xx})\,\mathbf{i} + \epsilon_{yx}\,\mathbf{j} + \epsilon_{zx}\,\mathbf{k}$$

$$\mathbf{j}^* = \epsilon_{xy}\,\mathbf{i} + (1 + \epsilon_{yy})\,\mathbf{j} + \epsilon_{zy}\,\mathbf{k}$$

$$\mathbf{k}^* = \epsilon_{xz}\,\mathbf{i} + \epsilon_{yz}\,\mathbf{j} + (1 + \epsilon_{zz})\,\mathbf{k}$$

The volume V* of the parallelepiped is given by the triple mixed product:

$$V^* = \mathbf{i}^* \cdot (\mathbf{j}^* \wedge \mathbf{k}^*)$$

In developing this product and retaining only the constant and the first-order terms of the coefficients of deformation, we obtain

$$V^* = 1 + \epsilon_{xx} + \epsilon_{yy} + \epsilon_{zz}$$

The ***quotient of the change of volume***, thus, is

$$\frac{V^* - 1}{1} = \epsilon_{xx} + \epsilon_{yy} + \epsilon_{zz} \tag{7.40}$$

The right side of relation (7.40) is the divergence of the displacement vector **F**, so that

$$\frac{V^* - 1}{1} = \text{div } \mathbf{F} \tag{7.41}$$

If the body is incompressible, the quotient of the change of volume cancels out, and we have as the ***condition of incompressibility***:

$$\text{div } \mathbf{F} = 0$$

7.5.4 Example: Rotation of a rigid body

During the rotation of a rigid body about a fixed line passing through the origin of the xyz coordinate system, the velocity vector of any point of the body is expressed by the formula:

$$\mathbf{v} = \boldsymbol{\omega} \wedge \mathbf{r},$$

where $\boldsymbol{\omega}$ is the angular velocity vector and $\mathbf{r} = x\mathbf{i} + y\mathbf{j} + z\mathbf{k}$ is the position vector of the point under consideration. Let s,t, and u be the components of the vector $\boldsymbol{\omega}$, which is assumed to be constant. The components of the vector product $\boldsymbol{\omega} \wedge \mathbf{r}$ are $tz - uy$, $ux - sz$, and $sy - tx$, so that the components of **curl** $(\boldsymbol{\omega} \wedge \mathbf{r})$ become 2s, 2t, and 2u. Thus, the angular velocity vector $\boldsymbol{\omega}$ is expressed as a function of the velocity vector **v** as follows:

$$\boldsymbol{\omega} = \frac{1}{2} \textbf{curl } (\boldsymbol{\omega} \wedge \mathbf{r}) = \frac{1}{2} \textbf{curl } \mathbf{v} \tag{7.42}$$

From this result, we see that the curl of a vector field is a measure of its angular rotation about each point.

7.5.5 Example: Wave equation

The propagation of an electromagnetic wave is described by the electrical field vector **E** (x,y,z,t,) and the magnetic field vector **H** (x,y,z,t). In a vacuum, the two field vectors **E** and **H** satisfy Maxwell's equations:

$$\textbf{curl } \mathbf{E} = -\mu_0 \frac{\partial \mathbf{H}}{\partial t} \tag{7.43}$$

$$\textbf{curl } \mathbf{H} = \epsilon_0 \frac{\partial \mathbf{E}}{\partial t} \tag{7.44}$$

and

$$\text{div } \mathbf{E} = \text{div } \mathbf{H} = 0 \tag{7.45}$$

where ϵ_0 and μ_0 are the electromagnetic characteristics of the vacuum. By taking the curl of the left and right sides of equation (7.43), we obtain, after a change in the order of differentiation, using relation (7.44):

$$\textbf{curl curl } \mathbf{E} = -\mu_0 \textbf{ curl}\left(\frac{\partial \mathbf{H}}{\partial t}\right) = -\mu_0 \frac{\partial}{\partial t}(\textbf{curl } \mathbf{H}) = -\epsilon_0\mu_0 \frac{\partial^2 \mathbf{E}}{\partial t^2}$$

According to the identity (7.36), we have

$$\textbf{curl curl } \mathbf{E} = -\Delta \mathbf{E} + \textbf{grad} \text{ div } \mathbf{E} = -\Delta \mathbf{E}$$

from which we get the ***wave equation***:

$$\Delta \mathbf{E} = \epsilon_0 \mu_0 \frac{\partial^2 \mathbf{E}}{\partial t^2} \tag{7.46}$$

In an analogous fashion, it can be shown that the magnetic field vector **H** satisfies the same differential equation:

$$\Delta \mathbf{H} = \epsilon_0 \mu_0 \frac{\partial^2 \mathbf{H}}{\partial t^2} \tag{7.47}$$

7.5.6 Example: Laplace's equation

Consider a scalar field ϕ (x,y,z). The equation:

$$\frac{\partial^2 \phi}{\partial x^2} + \frac{\partial^2 \phi}{\partial y^2} + \frac{\partial^2 \phi}{\partial z^2} = 0$$

or

$$\Delta \phi = 0 \tag{7.48}$$

is called ***Laplace's equation***.

We wish to determine the solutions ϕ of this equation, which depend only on the magnitude r of the position vector $\mathbf{r} = x\mathbf{i} + y\mathbf{j} + z\mathbf{k}$:

$$r = \sqrt{x^2 + y^2 + z^2}$$

The gradient of ϕ is by definition

$$\textbf{grad}\ \phi = \frac{\partial \phi}{\partial x}\mathbf{i} + \frac{\partial \phi}{\partial y}\mathbf{j} + \frac{\partial \phi}{\partial z}\mathbf{k}$$

and since it is assumed that $\phi = \phi\ (r)$, the first component, for example, of **grad** ϕ becomes

$$\frac{\partial \phi}{\partial x} = \frac{d\phi}{dr}\frac{\partial r}{\partial x} = \frac{d\phi}{dr}\frac{x}{\sqrt{x^2 + y^2 + z^2}} = \frac{d\phi}{dr}\frac{x}{r}$$

Thus, we obtain

$$\textbf{grad}\ \phi = \frac{1}{r}\frac{d\phi}{dr}\mathbf{r} \tag{7.49}$$

Because

$$\Delta\ \phi = \text{div}\ \textbf{grad}\ \phi = \text{div}\left(\frac{1}{r}\frac{d\phi}{dr}\mathbf{r}\right)$$

we find, by applying formula (7.33):

$$\Delta\phi = \frac{1}{r}\frac{d\phi}{dr}\ \text{div}\ \mathbf{r} + \mathbf{r}\cdot\textbf{grad}\left(\frac{1}{r}\frac{d\phi}{dr}\right) \tag{7.50}$$

For the divergence of **r**, we calculate

$$\text{div}\ \mathbf{r} = 3$$

and again using formula (7.49) for the gradient of a function that depends only on the magnitude r, we have

$$\textbf{grad}\left(\frac{1}{r}\frac{d\phi}{dr}\right) = \frac{1}{r}\frac{d}{dr}\left(\frac{1}{r}\frac{d\phi}{dr}\right)\mathbf{r}$$

and relation (7.50) becomes

$$\Delta\phi = \frac{3}{r}\frac{d\phi}{dr} + r\frac{d}{dr}\left(\frac{1}{r}\frac{d\phi}{dr}\right) \tag{7.51}$$

In this last algebraic transformation we have used the fact that $\mathbf{r}\cdot\mathbf{r} = \mathbf{r}^2$. Using (7.51), equation (7.48) takes the form:

$$\Delta\phi = \frac{2}{r}\frac{d\phi}{dr} + \frac{d^2\phi}{dr^2} = 0$$

which constitutes a first-order differential equation for $d\phi/dr$. Integrating this last equation, we find

$$\frac{d\phi}{dr} = \frac{c_1}{r^2}$$

and after a second integration:

$$\phi\,(r) = -\frac{c_1}{r}\;c_2 \tag{7.52}$$

where c_1 and c_2 are integration constants.

7.5.7 Example: Normal to a surface

Let us consider the surface described by the implicit equation $\phi\,(x,y,z) = c$, where c is a constant. Let $\mathbf{r} = x\mathbf{i} + y\mathbf{j} + z\mathbf{k}$ be the position vector of a point P (x,y,z) of the surface. Thus, the differential vector $\mathbf{dr} = dx\mathbf{i} + dy\mathbf{j} + dz\mathbf{k}$ is contained in the plane tangent to the surface at P. Because

$$d\phi = \frac{\partial\phi}{\partial x}\,dx + \frac{\partial\phi}{\partial y}\,dy + \frac{\partial\phi}{\partial z}\,dz$$

we have

$$\mathbf{grad}\;\phi \cdot \mathbf{dr} = 0 \tag{7.53}$$

i.e, **grad** ϕ is a vector perpendicular to **dr**, and thus to the surface, since **dr** is an arbitrary differential vector in the tangent plane.

Chapter 8

Space Curves and Line Integrals

8.1 SPACE CURVES

Let **r** (u) be a vector that depends on the parameter u. If **r** (u) is the position vector with end point P (x,y,z) in an xyz coordinate system, then

$$\mathbf{r}(u) = x(u)\,\mathbf{i} + y(u)\,\mathbf{j} + z(u)\,\mathbf{k}$$

As the parameter u changes, the end point of the vector **r** describes a ***space curve***.

Let us assume that the components of **r** (u) are continously differentiable, and that $d\mathbf{r}/du \neq \mathbf{0}$.

Then, the derivative

$$\frac{d\mathbf{r}}{du} = \lim_{\Delta u \to 0} \frac{\Delta \mathbf{r}}{\Delta u} = \lim_{\Delta u \to 0} \frac{\mathbf{r}(u+\Delta u) - \mathbf{r}(u)}{\Delta u} = \frac{dx}{du}\mathbf{i} + \frac{dy}{du}\mathbf{j} + \frac{dz}{du}\mathbf{k} \quad (8.1)$$

is a vector in the direction of the ***tangent*** to the curve in P (fig. 8.1).

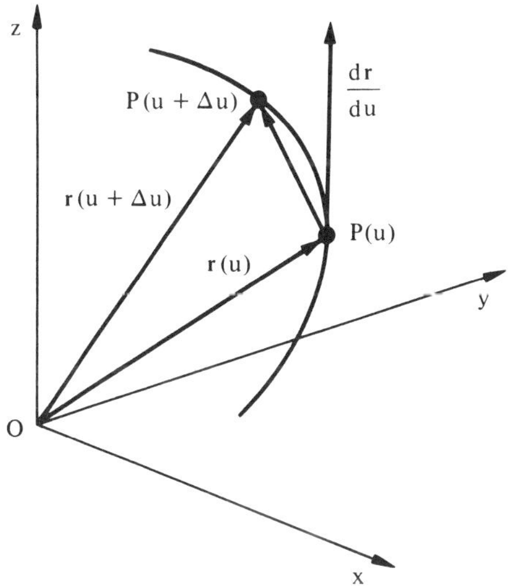

Fig. 8.1

As an example, we shall consider a ***circular helix*** of radius 2 coiling about the z-axis given by the vector function:

$$\mathbf{r}(\varphi) = 2\cos\varphi\,\mathbf{i} + 2\sin\varphi\,\mathbf{j} + \varphi/2\pi\,\mathbf{k}$$

For the tangent vector to the curve in $P(2\cos\varphi, 2\sin\varphi, \varphi/2\pi)$, we find:

$$\frac{d\mathbf{r}}{d\varphi} = -2\sin\varphi\mathbf{i} + 2\cos\varphi\mathbf{j} + \frac{1}{2\pi}\mathbf{k}$$

8.2 LENGTH OF A SPACE CURVE

Consider the space curve C, given by the vector function:

$$\mathbf{r} = \mathbf{r}(u), \qquad a \leq u \leq b$$

We consider a subdivision of the parameter interval $a \leq u \leq b$:

$$a = u_0 < u_1 < \ldots < u_n = b$$

This determines a sequence of points $P_0, P_1, \ldots, P_n$ on the arc C, corresponding to $\mathbf{r}_0 = \mathbf{r}(u_0)$, $\mathbf{r}_1 = \mathbf{r}(u_1), \ldots, \mathbf{r}_n = \mathbf{r}(u_n)$. Joining these points by a broken line of n chords, we obtain a polygon $P_0, P_1 \ldots, P_n$ inscribed to the arc C (fig. 8.2). The length l^* of this broken line of n chords equals

$$l^* = \sum_{i=0}^{n-1} |\mathbf{r}_{i+1} - \mathbf{r}_i| \qquad (8.2)$$

According to the mean-value theorem, we have

$$\mathbf{r}(u_{i+1}) - \mathbf{r}(u_i) = \frac{d\mathbf{r}}{du}(\xi_i)(u_{i+1} - u_i)$$

where $u_i \leq \xi_i \leq u_{i+1}$, and relation (8.2) becomes

$$l^* = \sum_{i=0}^{n-1} \left| \frac{d\mathbf{r}}{du}(\xi_i) \right| |u_{i+1} - u_i| \qquad (8.3)$$

The ***arc length*** of C is the limit of the sum (8.3) as the maximum chord length approaches zero as n approaches infinity, and is given by the integral:

$$l = \int_a^b \left| \frac{d\mathbf{r}}{du} \right| du = \int_a^b \sqrt{\left(\frac{dx}{du}\right)^2 + \left(\frac{dy}{du}\right)^2 + \left(\frac{dz}{du}\right)^2}\, du \qquad (8.4)$$

Going back to the example of the circular helix (section 8.1), we shall determine the length of one turn of the helix, i.e., the length of the arc:

$$\mathbf{r}(\varphi) = 2\cos\varphi\mathbf{i} + 2\sin\varphi\mathbf{j} + \varphi/2\pi\mathbf{k}, \quad 0 \leq \varphi \leq 2\pi$$

We calculate

$$l = \int_0^{2\pi} \sqrt{4 \sin^2 \varphi + 4 \cos^2 \varphi + \frac{1}{4\pi^2}}\, d\varphi = 2\pi \sqrt{4 + \frac{1}{4\pi^2}}$$

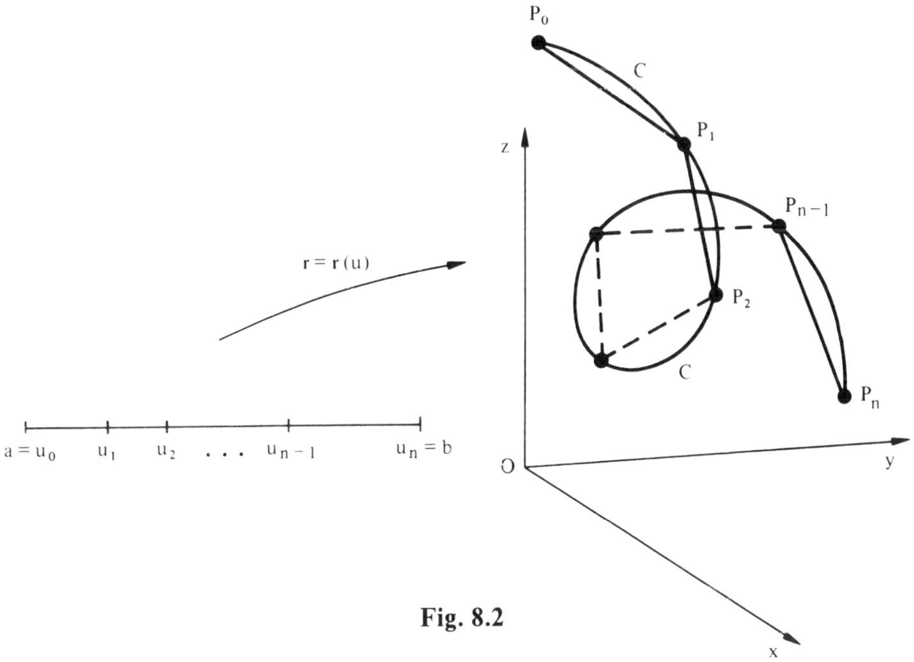

Fig. 8.2

8.3 ARC LENGTH

Let us again consider formula (8.4) for the length of a curve. If we replace the fixed upper limit of integration with a variable upper limit designated by u:

$$b = u$$

the length l of the curve will be a function of u:

$$l(u) = \int_a^u \sqrt{\left(\frac{dx}{du}\right)^2 + \left(\frac{dy}{du}\right)^2 + \left(\frac{dz}{du}\right)^2}\, du \tag{8.5}$$

This function l is called the ***arc length*** of C, and it will be designated hereafter by s(u).

Differentiating the integral (8.5) with respect to its upper limit, we find for the differential ds of the arc length s(u):

$$ds = \sqrt{\left(\frac{dx}{du}\right)^2 + \left(\frac{dy}{du}\right)^2 + \left(\frac{dz}{du}\right)^2}\, du \tag{8.6}$$

In the following, it will often be assumed that a space curve can be given as a function of its arc length s:

$$\mathbf{r}(s) = x(s)\mathbf{i} + y(s)\mathbf{j} + z(s)\mathbf{k}$$

It is clear that the representation of a space curve using the arc length as parameter is not unique; it depends on the choice of the initial point on the curve for which $s = 0$, and on the direction of increasing s. However, this representation has great advantages, for example,

$$\left|\frac{d\mathbf{r}}{ds}\right| = \left|\frac{d\mathbf{r}}{du}\right| \frac{1}{\sqrt{\left(\frac{dx}{du}\right)^2 + \left(\frac{dy}{du}\right)^2 + \left(\frac{dz}{du}\right)^2}} = 1 \tag{8.7}$$

i.e., the vector $d\mathbf{r}/ds$ is a ***unit*** vector in the direction of the tangent at any point of the curve.

8.4 ELEMENTS OF DIFFERENTIAL GEOMETRY

Let $\mathbf{r} = \mathbf{r}(s)$ be the position vector of a space curve whose parameter is the arc length s. We have just introduced the unit tangent vector $\mathbf{t} = d\mathbf{r}/ds$ (8.7). Because

$$\mathbf{t} \cdot \mathbf{t} = 1$$

we obtain, upon differentiating this equality with respect to the parameter s:

$$\frac{d\mathbf{t}}{ds} \cdot \mathbf{t} + \mathbf{t} \cdot \frac{d\mathbf{t}}{ds} = 2\mathbf{t} \cdot \frac{d\mathbf{t}}{ds} = 0$$

i.e., the derivative $d\mathbf{t}/ds$ of the unit tangent vector $\mathbf{t}$ is a vector perpendicular to $\mathbf{t}$; it is not generally a unit vector. The ***principal normal vector*** $\mathbf{n}$ is by definition the unit vector parallel to $d\mathbf{t}/ds$:

$$\mathbf{n} = \frac{d\mathbf{t}/ds}{|d\mathbf{t}/ds|} \tag{8.8}$$

The vector $d\mathbf{t}/ds$ indicates how fast the direction of the tangent is changing, and this is why its magnitude is called the ***curvature***. The reciprocal of the curvature $1/|d\mathbf{t}/ds|$ is designated by ρ and called the ***radius of curvature***. From (8.8), we have

$$\frac{d\mathbf{t}}{ds} = \frac{1}{\rho}\mathbf{n} \tag{8.9}$$

For a space curve, the curvature is by definition ≥ 0. We can easily see that the curvature of a space curve cancels identically if and only if the curve is a straight line. At any point of the space curve, we now have two vectors at right angles to each other. This enables us to set up a local coordinate system by introducing the unit vector **b** perpendicular to **t** and **n**:

$$\mathbf{b} = \mathbf{t} \wedge \mathbf{n} \tag{8.10}$$

which is called the ***binormal vector***. The vectors **t**, **n**, and **b** therefore, determine a moving ***trihedron*** at any point of the space curve. (fig. 8.3). The plane determined at the end point of the position vector by **t** and **n** is called the ***osculating*** plane, that determined by **n** and **b** the ***normal*** plane, and that by **t** and **b** the ***rectifying*** plane.

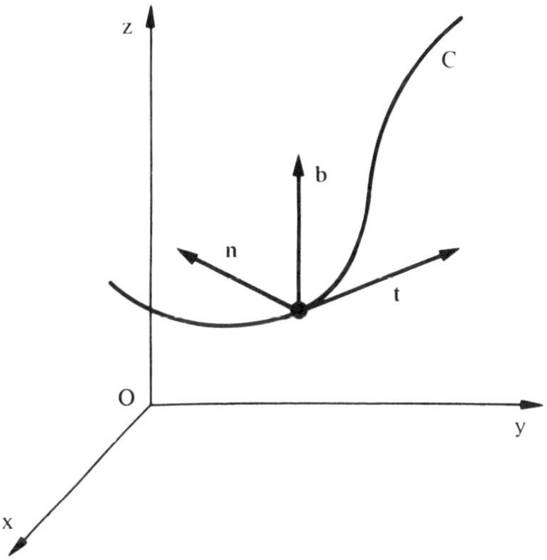

Fig. 8.3

The derivative $d\mathbf{b}/ds$ indicates how fast the direction of the binormal vector is changing, and this is why its magnitude is called the ***torsion***. From (8.10), we have

$$\frac{d\mathbf{b}}{ds} = \frac{d\mathbf{t}}{ds} \wedge \mathbf{n} + \mathbf{t} \wedge \frac{d\mathbf{n}}{ds} = \mathbf{t} \wedge \frac{d\mathbf{n}}{ds} \tag{8.11}$$

since the vectors $d\mathbf{t}/ds$ and **n** are parallel. The vector $d\mathbf{b}/ds$ is, therefore, perpendicular to **t**. On the other hand, the derivative $d\mathbf{b}/ds$ is perpendicular to the vector

b, since **b** is a unit vector. Thus, the vector $d\mathbf{b}/ds$ is perpendicular to the vectors **t** and **b**, i.e., it must be parallel to the principal normal vector **n**. Consequently,

$$\frac{d\mathbf{b}}{ds} = -\frac{1}{\tau}\mathbf{n} \tag{8.12}$$

The length of the scalar τ is the ***torsion radius*** and $1/\tau$ is the ***torsion***. The negative sign in (8.12) is introduced so that the torsion is positive when the vector triad rotates in a right-handed sense about the tangent as it moves along the curve. We can also easily see that the torsion of a curve cancels identically if and only if the curve is a plane curve.

Because **t**, **n**, and **b** form a right-handed triad ($\mathbf{b} = \mathbf{t} \wedge \mathbf{n}$), it is the same for **b**, **t**, and **n**, i.e.,

$$\mathbf{n} = \mathbf{b} \wedge \mathbf{t}$$

Thus, by differentiating the vector **n** with respect to s:

$$\frac{d\mathbf{n}}{ds} = \frac{d\mathbf{b}}{ds} \wedge \mathbf{t} + \mathbf{b} \wedge \frac{d\mathbf{t}}{ds}$$

and using definitions (8.9) and (8.12), we obtain

$$\frac{d\mathbf{n}}{ds} = -\frac{1}{\tau}(\mathbf{n} \wedge \mathbf{t}) + \frac{1}{\rho}(\mathbf{b} \wedge \mathbf{n})$$

from which we get

$$\frac{d\mathbf{n}}{ds} = \frac{1}{\tau}\mathbf{b} - \frac{1}{\rho}\mathbf{t} \tag{8.13}$$

Equations (8.9), (8.12), and (8.13) are the famous ***Frenet-Serret formulas***.

8.4.1 Example: Differential geometry of the helix

Consider the circular helix of radius 3 about the z-axis (8.4), given by

$$\mathbf{r}(\varphi) = 3 \cos \varphi \mathbf{i} + 3 \sin \varphi \mathbf{j} + 4 \varphi \mathbf{k} \tag{8.14}$$

According to formula (8.5), we calculate for its arc length s from the point A(1, 0, 0), in which $\varphi = 0$:

$$s = \int_0^{\varphi} \sqrt{\left(\frac{dx}{d\varphi}\right)^2 + \left(\frac{dy}{d\varphi}\right)^2 + \left(\frac{dz}{d\varphi}\right)^2}\, d\varphi = 5\varphi \tag{8.15}$$

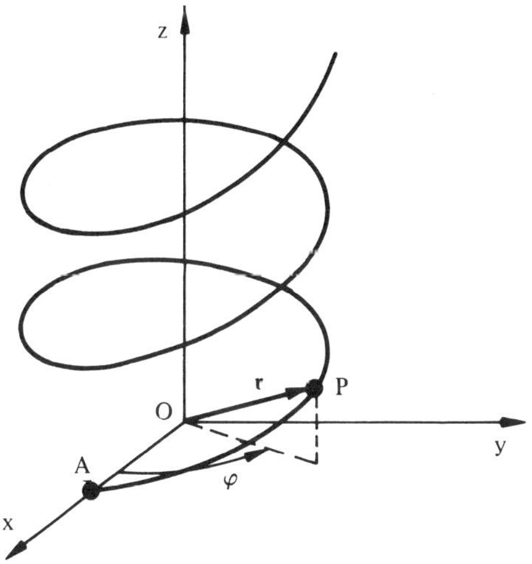

Fig. 8.4

and the representation of the helix (8.14) as a function of its arc length is the following:

$$\mathbf{r}(s) = 3 \cos \frac{s}{5}\, \mathbf{i} + 3 \sin \frac{s}{5}\, \mathbf{j} + \frac{4}{5}\, s\, \mathbf{k}$$

From relation (8.15), the differential ds becomes

$$ds = 5\, d\varphi$$

Thus, for the unit tangent vector **t**:

$$\mathbf{t} = \frac{d\mathbf{r}}{ds} = \frac{d\mathbf{r}}{d\varphi}\frac{d\varphi}{ds} = \frac{1}{5}(-3 \sin \varphi\, \mathbf{i} + 3 \cos \varphi\, \mathbf{j} + 4\, \mathbf{k})$$

and

$$\frac{d\mathbf{t}}{ds} = \frac{d\mathbf{t}}{d\varphi}\frac{d\varphi}{ds} = \frac{1}{25}(-3 \cos \varphi\, \mathbf{i} - 3 \sin \varphi\, \mathbf{j})$$

According to formula (8.9):

$$\frac{d\mathbf{t}}{ds} = \frac{1}{\rho}\mathbf{n} \text{ (ρ is positive by definition)}$$

we find for the radius of curvature ρ:

$$\rho = 25/3$$

The principal normal vector **n** is given by

$$\mathbf{n} = -(\cos\varphi\,\mathbf{i} + \sin\varphi\,\mathbf{j})$$

such that

$$\mathbf{b} = \mathbf{t} \wedge \mathbf{n} = \frac{4}{5}\sin\varphi\,\mathbf{i} - \frac{4}{5}\cos\varphi\,\mathbf{j} + \frac{3}{5}\mathbf{k}$$

It follows that

$$\frac{d\mathbf{b}}{ds} = \frac{d\mathbf{b}}{d\varphi}\frac{d\varphi}{ds} = \frac{4}{25}\cos\varphi\,\mathbf{i} + \frac{4}{25}\sin\varphi\,\mathbf{j}$$

and from formula (8.12):

$$\frac{d\mathbf{b}}{ds} = -\frac{1}{\tau}\mathbf{n}$$

we obtain for the radius of torsion τ:

$$\tau = 25/4$$

8.4.2 Example: Geometric interpretation of the radius of curvature

We can interpret the curvature $1/\rho$ (8.9) in the following way. Let us consider the principal normal vectors of a space curve at a point P(s) and at a neighboring point P(s+Δs): **n**(s) and **n**(s+Δs), and let us designate by **R** the point of intersection of the vectors **n**(s) and **n**(s+Δs). Let ρ^* be the length of the segment $\overline{P(s)R}$ and $\Delta\alpha$ the angle between **n**(s) and **n**(s + Δs) (fig. 8.5). We obtain at the limit as $\Delta s \to 0$:

$$\lim_{\Delta s \to 0}\frac{1}{\rho^*} = \lim_{\Delta s \to 0}\frac{\Delta\alpha}{\Delta s} = \lim_{\Delta s \to 0}\frac{\Delta\alpha}{|\Delta\mathbf{t}|}\left|\frac{\Delta\mathbf{t}}{\Delta s}\right| \tag{8.16}$$

Because **t** is a unit vector, we have (fig. 8.6)

$$\lim_{\Delta s \to 0}\frac{\Delta\alpha}{|\Delta\mathbf{t}|} = 1$$

and equation (8.16) becomes

$$\lim_{\Delta s \to 0}\frac{1}{\rho^*} = \lim_{\Delta s \to 0}\frac{\Delta\alpha}{|\Delta\mathbf{t}|}\lim_{\Delta s \to 0}\left|\frac{\Delta\mathbf{t}}{\Delta s}\right| = \lim_{\Delta s \to 0}\left|\frac{\Delta\mathbf{t}}{\Delta s}\right| = \frac{1}{\rho}$$

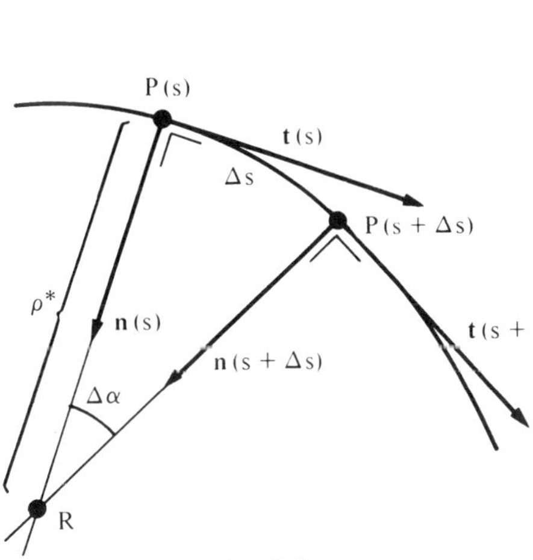

Fig. 8.5

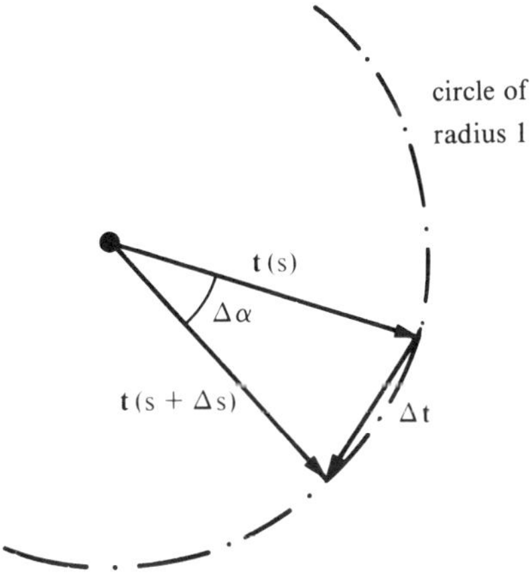

Fig. 8.6

8.5 CURVILINEAR MOVEMENT

A particle moves along a space curve described by the position vector:

$$\mathbf{r}(t) = x(t)\mathbf{i} + y(t)\mathbf{j} + z(t)\mathbf{k} \tag{8.17}$$

where the parameter t designates the time.

The ***velocity vector*** $\mathbf{v}$ of the moving point P is equal to the ***first derivative of the position vector*** $\mathbf{r}$ with respect to time:

$$\mathbf{v} = \frac{d\mathbf{r}}{dt} \tag{8.18}$$

The magnitude $|\mathbf{v}|$ of the velocity vector $\mathbf{v}$ is the ***speed*** of the point P:

$$v = |\mathbf{v}| = \sqrt{\left(\frac{dx}{dt}\right)^2 + \left(\frac{dy}{dt}\right)^2 + \left(\frac{dz}{dt}\right)^2} = \frac{ds}{dt} \tag{8.19}$$

If the arc length s is introduced as a parameter into (8.17), we will have for the velocity vector $\mathbf{v}$:

$$\mathbf{v} = \frac{d\mathbf{r}}{dt} = \frac{d\mathbf{r}}{ds}\frac{ds}{dt} = \mathbf{t}\, v \tag{8.20}$$

where $\mathbf{t}$ is the unit vector tangent to the curve (8.17) pointing in the direction of the movement.

The ***acceleration vector*** $\mathbf{a}$ is the ***derivative of the velocity vector*** $\mathbf{v}$ with respect to time:

$$\mathbf{a} = \frac{d\mathbf{v}}{dt} \tag{8.21}$$

Since $\mathbf{v} = d\mathbf{r}/dt$, we have

$$\mathbf{a} = \frac{d^2\mathbf{r}}{dt^2} = \frac{d}{dt}(\mathbf{t}\,v) = \frac{d\mathbf{t}}{dt}v + \mathbf{t}\frac{dv}{dt}$$

and

$$\frac{d\mathbf{t}}{dt} = \frac{d\mathbf{t}}{ds}\frac{ds}{dt} = \frac{1}{\rho}\mathbf{n}\,v$$

by using formula (8.9). Thus, we obtain for the acceleration vector **a**:

$$\mathbf{a} = \frac{dv}{dt}\mathbf{t} + \frac{v^2}{\rho}\mathbf{n} \tag{8.22}$$

where **t** is the unit tangent vector, always oriented in the direction of the movement, **n** is the principal normal vector, and ρ is the radius of curvature of the trajectory (8.17).

8.5.1 Example: Ball sliding on a cylinder

Let us consider a ball of mass m placed at the top of a cylinder of radius R and with a horizontal axis. This position is unstable, and at time t = 0, the ball starts to slide due to its weight. Neglecting friction, we wish to find the ***point of separation*** of the ball from the cylinder (fig. 8.7).

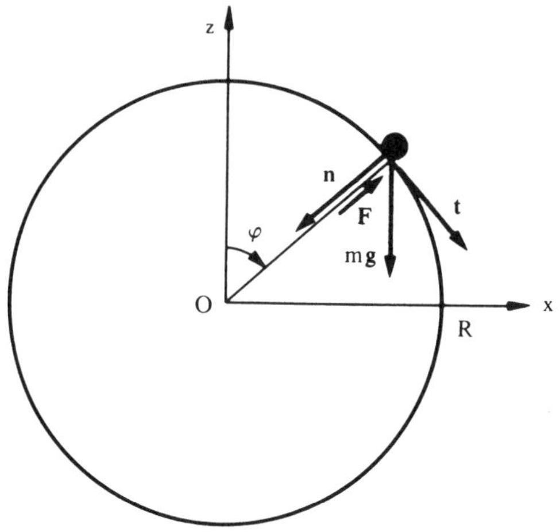

Fig. 8.7

As the weight $m\mathbf{g}$ and the reaction $\mathbf{F}$ of the cylinder take effect in a vertical plane, the movement of the ball will also take place in the same plane. If φ designates the angle between the position vector of the ball and the z-axis, taken in the negative direction, the length s of the path taken by the ball becomes

$$s = R\,\varphi$$

from which we get the speed v:

$$v = \frac{ds}{dt} = R\,\dot{\varphi}$$

and for its derivative:

$$\frac{dv}{dt} = R\,\ddot{\varphi}$$

The movement of the ball is governed by ***Newton's law***:

$$m\,\frac{d^2\mathbf{r}}{dt^2} = m\mathbf{g} + \mathbf{F}$$

This last vector equation can be expressed in the moving coordinate system spanned by the orthogonal unit vectors **t** and **n**. According to formula (8.2), the tangential component becomes

$$m\,\frac{dv}{dt} = mg\sin\varphi \tag{8.23}$$

and the normal component:

$$\frac{mv^2}{R} = mg\cos\varphi - F \tag{8.24}$$

where $F = |\mathbf{F}|$. The point of separation is characterized by the constraint $F = 0$.

Equation (8.23) states that

$$m\,\frac{dv}{dt} = m\,R\,\ddot{\varphi} = mg\sin\varphi$$

from which we get

$$R\,\ddot{\varphi} = g\sin\varphi \tag{8.25}$$

Let us assume that $\dot{\varphi}$ depends only on φ:

$$\dot{\varphi} = f(\varphi)$$

Thus, we have for the second derivative $\ddot{\varphi}$:

$$\ddot{\varphi} = f' \dot{\varphi} = f' f$$

and substituting this relation into equation (8.25), we find

$$R f f' = g \sin \varphi$$

Separating the variables, we have

$$df \cdot f = \frac{g}{R} \sin \varphi \, d \varphi$$

and integrating this last equation, we have

$$\frac{f^2}{2} = -\frac{g}{R} \cos \varphi + c$$

Since $f = \dot{\varphi}$, we obtain

$$\dot{\varphi}^2 = -\frac{2g}{R} \cos \varphi + 2c \tag{8.26}$$

The initial condition $\dot{\varphi} = 0$ for $\varphi = 0$ gives for the integration constant $c = g/R$, and equation (8.26) becomes

$$\dot{\varphi}^2 = -\frac{2g}{R} (1 - \cos \varphi) \tag{8.27}$$

Taking the condition of separation $F = 0$ into account, equation (8.24) takes the form:

$$\frac{mv^2}{R} = m R \dot{\varphi}^2 = mg \cos \varphi$$

and substituting $\dot{\varphi}^2$ with the aid of equation (8.27), we find the following equation in φ:

$$2(1 - \cos \varphi) = \cos \varphi$$

from which we get

$$\cos \varphi = 2/3$$

and

$$\varphi = 48°$$

8.6 LINE INTEGRALS

Let us consider the arc of a space curve C that links the points P_0 and P_1 given by the position vector:

$\mathbf{r} = \mathbf{r}(u), \quad a \leq u \leq b$

In addition, let **F** represent a vector field defined at each point of C:

$$\mathbf{F} = \mathbf{F}(x, y, z) = F_x(x, y, z)\mathbf{i} + F_y(x, y, z)\mathbf{j} + F_z(x, y, z)\mathbf{k}$$

The ***line integral*** I of **F** along the space curve between the two specified points is defined by

$$I = \int_{P_0}^{P_1} \mathbf{F} \cdot d\mathbf{r} = \int_{P_0}^{P_1} F_x\, dx + F_y\, dy + F_z\, dz \tag{8.28}$$

For a physical interpretation of this integral, let us consider a particle **P** being displaced along C under the action of a force **F**. C is subdivided into n line segments by the points $P_0 = Q_0, Q_1, \ldots, Q_{n-1}, Q_n = P_1$. We designate the vector $Q_i Q_{i+1}$ by $\Delta\mathbf{r}_i$ and the force acting on point Q_i by $\mathbf{F}_i$ (fig. 8.8).

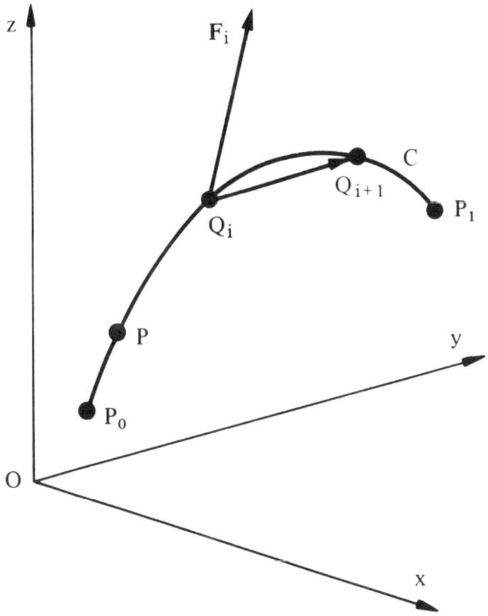

Fig. 8.8

The scalar product $\mathbf{F}_i \cdot \Delta\mathbf{r}_i$ approximates the work ΔT done by the force in moving the particle from Q_i to Q_{i+1}:

$$\Delta T \approx \mathbf{F}_i \cdot \Delta\mathbf{r}_i$$

and the sum approximates the total ***work*** T done on the particle when it is moved from P_0 to P_1:

$$T \approx \sum_{i=0}^{n-1} \mathbf{F}_i \cdot \Delta \mathbf{r}_i \tag{8.29}$$

As $|\Delta \mathbf{r}_i| \to 0$, the limit of the sum on the right side of (8.29) is the line integral of **F** along **C**, i.e.,

$$T = \lim_{|\Delta \mathbf{r}_i| \to 0} \sum \mathbf{F}_i \cdot \Delta \mathbf{r}_i = \int_{P_0}^{P_1} \mathbf{F} \cdot d\mathbf{r}$$

8.6.1 Example: Calculation of a line integral

As an example, we shall calculate the line integral of the vector field $\mathbf{F} = x^3\mathbf{i} + 3zy^2\mathbf{j} - x^2y\mathbf{k}$ along the straight-line segment that connects the point P_0 (3, 2, 1) to the point P_1 (0, 1, 0). This segment permits the following vector representation:

$$\mathbf{r}(u) = OP_0 + uP_0P_1 = (3 - 3u)\mathbf{i} + (2 - u)\mathbf{j} + (1 - u)\mathbf{k}, \quad 0 \leq u \leq 1$$

Applying formula (8.28), we calculate

$$\int_{P_0}^{P_1} \mathbf{F} \cdot d\mathbf{r} = \int_{P_0}^{P_1} x^3\, dx + 3zy^2\, dy - x^2y\, dz$$

$$= \int_0^1 [(3 - 3u)^3 (-3) + 3(1 - u)(2 - u)^2 (-1) - (3 - 3u)^2 (2 - u)(-1)]\, du$$

$$= \int_0^1 (75u^3 - 222u^2 + 222u - 75)\, du = -19.25$$

Chapter 9

Surfaces and Surface Integrals

9.1 SURFACES

In elementary analysis, a surface in three dimensions is either specified in ***explicit form***:

$$z = f(x, y) \tag{9.1}$$

or in ***implicit form:***

$$F(x, y, z) = 0 \tag{9.2}$$

For example, $x^2 + y^2 + z^2 - R^2 = 0$ is the implicit representation of a sphere of radius R centered at the origin, and $z = \pm\sqrt{R^2 - x^2 - y^2}$ is its explicit representation.

In a more elegant way, a surface can be represented in ***parametric form***, i.e., the coordinates of its points are functions of two parameters u and v:

$$x = x(u, v), \qquad y = y(u, v), \qquad z = z(u, v) \tag{9.3}$$

where u and v vary in a given domain D of the uv plane.

If **r** designates the position vector for a point P of the surface, representation (9.3) is equivalent to the following ***vector representation***:

$$\mathbf{r}(u, v) = x(u, v)\mathbf{i} + y(u, v)\mathbf{j} + z(u, v)\mathbf{k}, \qquad (u, v) \in D \tag{9.4}$$

and it is this representation which we will use hereafter.

A **surface** S is, thus, the image in space of the domain of variation D of the parameters u and v under the application $\mathbf{r} = \mathbf{r}(u,v)$ (fig. 9.1).

Let us assume that the x,y, and z components of the vector **r** are continuously differentiable with respect to u and v. Then, the image of the line $v = v_0$ = constant is the space curve $\mathbf{r} = \mathbf{r}(u,v_0)$ on the surface for which u is the parameter, which is called the u — ***coordinate line***. Likewise, the image of the line $u = u_0$ = constant is the curve $\mathbf{r} = \mathbf{r}(u_0,v)$ on the surface for which v is the parameter, which is called

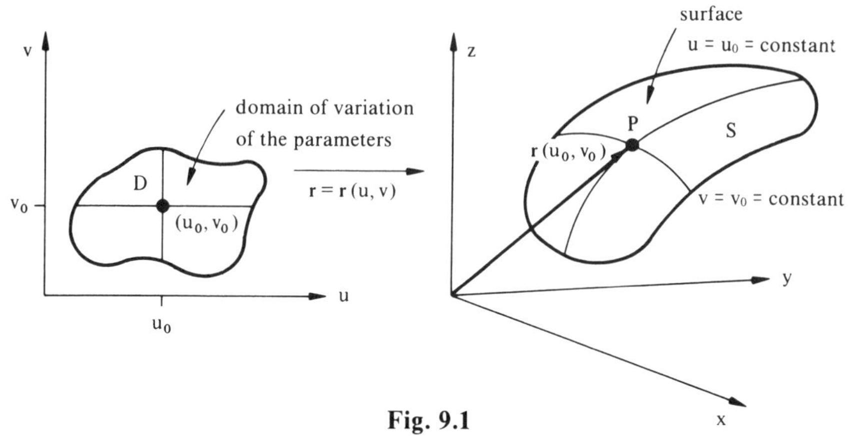

Fig. 9.1

the v — ***coordinate line***.. All the u and v coordinate lines form two families of curves which cover the surface.

As an example, we shall look at the sphere of radius R centered at the origin. It allows the following vector representation:

$$\mathbf{r}(\vartheta, \varphi) = R \sin \vartheta \cos \varphi\, \mathbf{i} + R \sin \vartheta \sin \varphi\, \mathbf{j} + R \cos \vartheta\, \mathbf{k}$$

where the domain of variation of the parameters ϑ and φ is determined by the inequalities $0 \leq \vartheta \leq \pi$ and $0 \leq \varphi \leq 2\pi$.

The φ-coordinate lines ($\vartheta = \vartheta_0$ = constant) are the ***latitude circles*** of the sphere and the ϑ-coordinate lines ($\varphi = \varphi_0$ = constant) are its ***meridians*** (fig. 9.2).

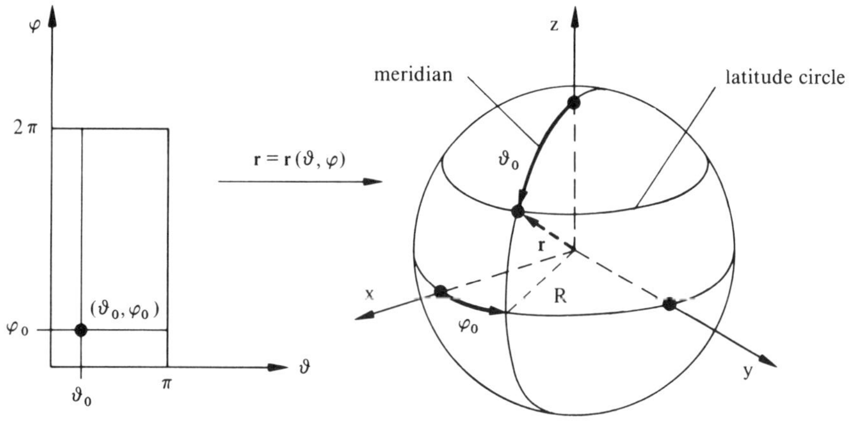

Fig. 9.2

9.2 ELEMENTS OF DIFFERENTIAL GEOMETRY

Consider the surface S, given in vector form:

$$\mathbf{r} = \mathbf{r}(u, v), \qquad (u, v) \in D$$

We are interested in the ***differential surface element*** $d\sigma$ of S. With this aim, we consider a small section of surface dS of S limited by two pairs of neighboring coordinate lines: v-coordinate lines, u and u + du, and u-coordinate lines, v and v + dv (fig. 9.3).

The vector $\partial\mathbf{r}/\partial u \, du$ is a vector tangent to the u-coordinate line, and $\partial\mathbf{r}/\partial v \, dv$ is tangent to the v-coordinate line. The vector product of these two tangent vectors is a vector normal to the surface whose length is equal to the area of the parallelogram whose sides are $\partial\mathbf{r}/\partial u \, du$ and $\partial\mathbf{r}/\partial v \, dv$ (fig. 9.4), and it is by means of this area that we approximate the area $d\sigma$ of dS.

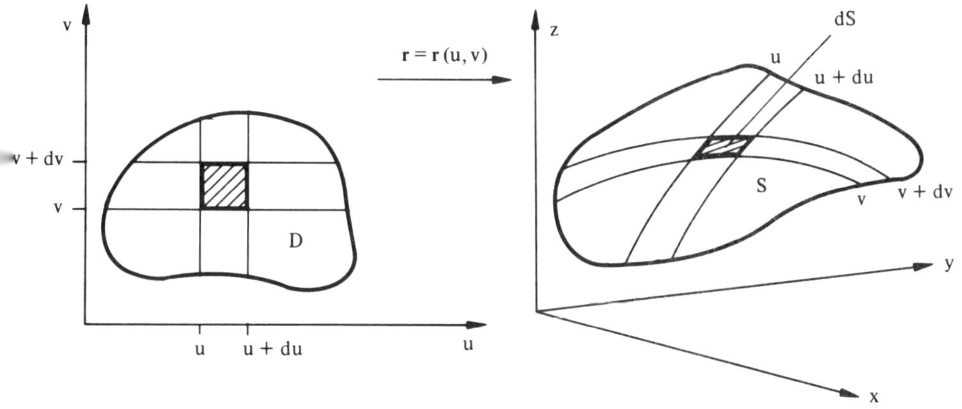

Fig. 9.3

If **n** designates the ***unit vector normal*** to the surface, we have

$$\left(\frac{\partial\mathbf{r}}{\partial u} \wedge \frac{\partial\mathbf{r}}{\partial v}\right) du\, dv = \mathbf{n}\, d\sigma \tag{9.5}$$

Assuming that $\partial\mathbf{r}/\partial u \wedge \partial\mathbf{r}/\partial v \neq \mathbf{0}$, we orient the vector **n** (without indications to the contrary) so that $\partial\mathbf{r}/\partial u$, $\partial\mathbf{r}/\partial v$ and **n** form a right-handed system.

Using the identity:

$$\left(\frac{\partial\mathbf{r}}{\partial u} \wedge \frac{\partial\mathbf{r}}{\partial v}\right)^2 = \left(\frac{\partial\mathbf{r}}{\partial u}\right)^2 \left(\frac{\partial\mathbf{r}}{\partial v}\right)^2 - \left(\frac{\partial\mathbf{r}}{\partial u} \cdot \frac{\partial\mathbf{r}}{\partial v}\right)^2$$

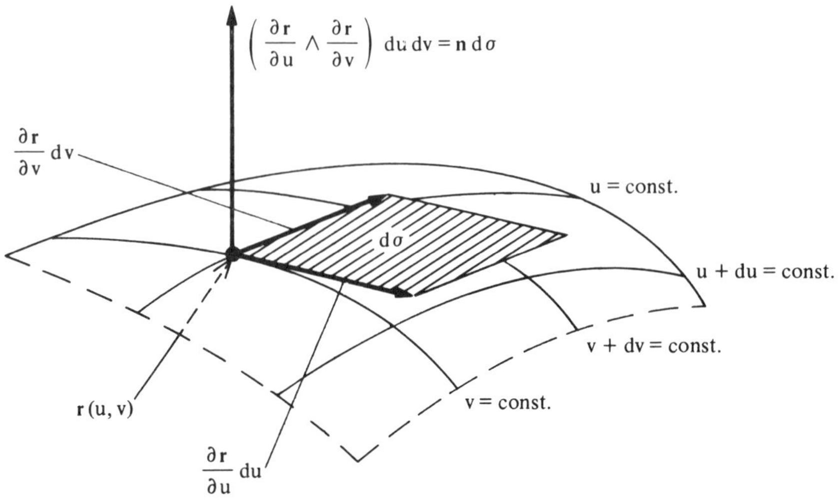

Fig. 9.4

and the definitions:

$$E = \left(\frac{\partial \mathbf{r}}{\partial u}\right)^2, \quad F = \frac{\partial \mathbf{r}}{\partial u} \cdot \frac{\partial \mathbf{r}}{\partial v}, \quad G = \left(\frac{\partial \mathbf{r}}{\partial v}\right)^2 \tag{9.6}$$

we obtain the following expression for the ***differential surface element*** dσ from equation (9.5):

$$d\sigma = \sqrt{EG - F^2}\ du\, dv \tag{9.7}$$

To obtain the differential ds of arc length, we take its square, getting

$$ds^2 = d\mathbf{r} \cdot d\mathbf{r} = \left(\frac{\partial \mathbf{r}}{\partial u} du + \frac{\partial \mathbf{r}}{\partial v} dv\right)^2$$

$$= \left(\frac{\partial \mathbf{r}}{\partial u}\right)^2 du^2 + 2\, \frac{\partial \mathbf{r}}{\partial u} \cdot \frac{\partial \mathbf{r}}{\partial v}\, du\, dv + \left(\frac{\partial \mathbf{r}}{\partial v}\right)^2 dv^2$$

$$= E\, du^2 + 2\, F\, du\, dv + g\, dv^2 \tag{9.8}$$

which constitutes the ***first fundamental form*** of the surface.

As an example, let us consider the sphere of radius R centered at the origin:

$$\mathbf{r}(\vartheta, \varphi) = R \sin \vartheta \cos \varphi \, \mathbf{i} + R \sin \vartheta \sin \varphi \, \mathbf{j} + R \cos \vartheta \, \mathbf{k}$$

$$0 \leq \vartheta \leq \pi, \quad 0 \leq \varphi \leq 2\pi$$

The derivatives of the position vector $\mathbf{r}$ with respect to the parameters ϑ and φ are:

$$\frac{\partial \mathbf{r}}{\partial \vartheta} = R \cos \vartheta \cos \varphi \, \mathbf{i} + R \cos \vartheta \sin \varphi \, \mathbf{j} - R \sin \vartheta \, \mathbf{k}$$

$$\frac{\partial \mathbf{r}}{\partial \varphi} = -R \sin \vartheta \sin \varphi \, \mathbf{i} + R \sin \vartheta \cos \varphi \, \mathbf{j}$$

and from formula (9.5) we calculate

$$\begin{aligned} \mathbf{n} \, d\sigma &= \left(\frac{\partial \mathbf{r}}{\partial \vartheta} \wedge \frac{\partial \mathbf{r}}{\partial \vartheta} \right) d\vartheta \, d\varphi \\ &= R^2 \sin \vartheta \, (\sin \vartheta \cos \varphi \, \mathbf{i} + \sin \vartheta \sin \varphi \, \mathbf{j} + \cos \vartheta \, \mathbf{k}) \, d\vartheta \, d\varphi \\ &= R \sin \vartheta \, \mathbf{r} \, d\vartheta \, d\varphi = R^2 \sin \vartheta \, \mathbf{n} \, d\vartheta \, d\varphi \end{aligned}$$

from which we get

$$d\sigma = R^2 \sin \vartheta \, d\vartheta \, d\varphi$$

The area Σ of the sphere of radius R is, thus, the double integral:

$$\Sigma = \int_{\varphi=0}^{2\pi} \int_{\vartheta=0}^{\pi} R^2 \sin \vartheta \, d\vartheta \, d\varphi = 4\pi R^2$$

The square of the differential of arc length ds^2 is given by the formula (9.8):

$$ds^2 = E \, d\vartheta^2 + 2 F \, d\vartheta \, d\varphi + G \, d\varphi^2$$

and since

$$E = \left(\frac{\partial \mathbf{r}}{\partial \vartheta} \right)^2 = R^2$$

$$F = \frac{\partial \mathbf{r}}{\partial \vartheta} \cdot \frac{\partial \mathbf{r}}{\partial \varphi} = 0$$

$$G = \left(\frac{\partial \mathbf{r}}{\partial \varphi} \right)^2 = R^2 \sin^2 \vartheta$$

we get

$$ds^2 = R^2\, d\vartheta^2 + R^2 \sin^2 \vartheta\, d\varphi^2$$

9.3 SURFACE INTEGRALS

Consider a surface S, given in vector form:

$$\mathbf{r} = \mathbf{r}(u, v), \quad (u, v) \in D$$

Let us assume that the vector field **F** is defined in each point of S, **F** being given in terms of its components:

$$\mathbf{F} = \mathbf{F}(x, y, z) = F_x(x, y, z)\mathbf{i} + F_y(x, y, z)\mathbf{j} + F_z(x, y, z)\mathbf{k}$$

We subdivide S into pieces of surface ΔS_i, and at each ΔS_i, we fix a point P_i. Then we consider the sum:

$$\sum_i \mathbf{F}_i \cdot \mathbf{n}_i\, \Delta \sigma_i \tag{9.9}$$

where $\mathbf{F}_i$ designates the vector **F** in P_i, $\mathbf{n}_i$ is the normal unit vector to the surface at this point, and $\Delta\sigma_i$ is the area of ΔS_i (fig. 9.5).

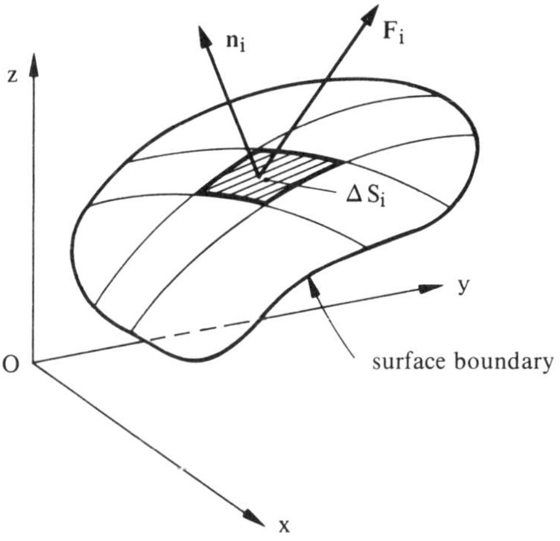

Fig. 9.5

The limit of the sum (9.9), when the diameter of the surface elements tends toward zero, is called the ***surface integral*** of thc vcctor function **F** over S. The following notation is used:

$$I = \lim_{\Delta \sigma_i \to 0} \sum \mathbf{F}_i \cdot \mathbf{n}_i \, \Delta \sigma_i = \iint_S \mathbf{F} \cdot \mathbf{n} \, d\sigma \tag{9.10}$$

It must be noted again that each term of the sum (9.9) is equal to the volume of the cylinder of base ΔS_i and height $\mathbf{F}_i \cdot \mathbf{n}_i$. If **F** indicates the velocity of a fluid crossing the surface S, the surface integral (9.10) is thus equal to the quantity of fluid crossing the surface per unit time.

The notation:

$$\oiint_S \mathbf{F} \cdot \mathbf{n} \, d\sigma$$

is used to indicate the integration over a ***closed surface*** (for example, over a sphere, an ellipsoid, *et cetera*).

9.3.1 Example: Calculation of a surface integral

Calculate the surface integral:

$$\iint_S \mathbf{F} \cdot \mathbf{n} \, d\sigma$$

where $\mathbf{F} = 18\,z\mathbf{i} - 12\mathbf{j} + 3\,y\mathbf{k}$ and S is part of the plane $2x + 3y + 6z = 12$ located in the first octant (fig. 9.6).

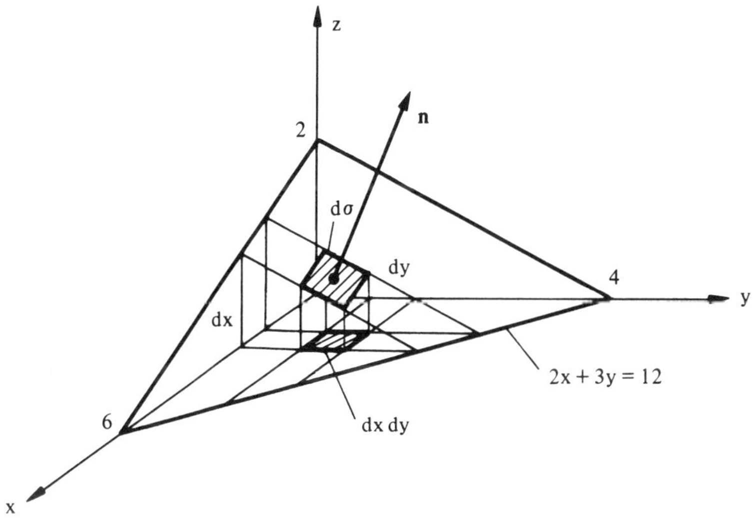

Fig. 9.6

Choosing x and y as parameters, the specified surface has the following vector representation:

$$\mathbf{r}(x, y) = x\mathbf{i} + y\mathbf{j} + \left(2 - \frac{1}{3}x - \frac{1}{2}y\right)\mathbf{k}$$

The domain of variation D of the parameters x and y is the projection of the surface on the xy plane, described by the inequalities $0 \leq y \leq 4 - 2/3\ x$ and $0 \leq x \leq 6$.

To determine $\mathbf{n}\, d\sigma$, we apply the formula (9.5):

$$\mathbf{n}\, d\sigma = \left(\frac{\partial \mathbf{r}}{\partial x} \wedge \frac{\partial \mathbf{r}}{\partial y}\right) dx\, dy$$

Because

$$\mathbf{r} = x\mathbf{i} + y\mathbf{j} + z\mathbf{k} = x\mathbf{i} + y\mathbf{j} + \left(2 - \frac{1}{3}x - \frac{1}{2}y\right)\mathbf{k}$$

we calculate

$$\frac{\partial \mathbf{r}}{\partial x} = \mathbf{i} - \frac{1}{3}\mathbf{k}$$

$$\frac{\partial \mathbf{r}}{\partial y} = \mathbf{j} - \frac{1}{2}\mathbf{k}$$

such that

$$\frac{\partial \mathbf{r}}{\partial x} \wedge \frac{\partial \mathbf{r}}{\partial y} = \frac{1}{3}\mathbf{i} + \frac{1}{2}\mathbf{j} + \mathbf{k}$$

and, thus,

$$\mathbf{n}\, d\sigma = \left(\frac{1}{3}\mathbf{i} + \frac{1}{2}\mathbf{j} + \mathbf{k}\right) dx\, dy$$

On the surface considered, we have

$$\mathbf{F} \cdot \mathbf{n}\, d\sigma = (6z - 6 + 3y)\, dx\, dy = (6 - 2x)\, dx\, dy$$

since

$$z = 2 - \frac{1}{3}x - \frac{1}{2}y$$

and the integral of **F** over the surface S takes the following form:

$$\iint_S \mathbf{F} \cdot \mathbf{n} \, d\sigma = \iint_D (6 - 2x) \, dx \, dy$$

For the evaluation of the last integral, we first fix x and integrate with respect to y from y = 0 to y = 4 – 2/3x. Then we integrate with respect to x from x = 0 to x = 6, which gives

$$\iint_S \mathbf{F} \cdot \mathbf{n} \, d\sigma = \int_{x=0}^{6} \left\{ \int_{y=0}^{4-\frac{2}{3}x} (6 - 2x) \, dy \right\} dx = 24$$

Chapter 10

Divergence Theorem, Gradient Theorem, and Green's Theorem

10.1 DIVERGENCE THEOREM

The ***divergence theorem*** states that if V is the volume bounded by a closed surface S and if **F** (x,y,z) is a vector field with continuous partial derivatives defined in the region V and on its boundary S then

$$\iiint_V \operatorname{div} \mathbf{F}\, dV = \oiint_S \mathbf{F} \cdot \mathbf{n}\, d\sigma \tag{10.1}$$

where **n** designates the normal unit vector directed toward the exterior of S. This fundamental formula gives a relation between the volume integral over V and the closed surface integral over S, where S limits V (fig. 10.1).

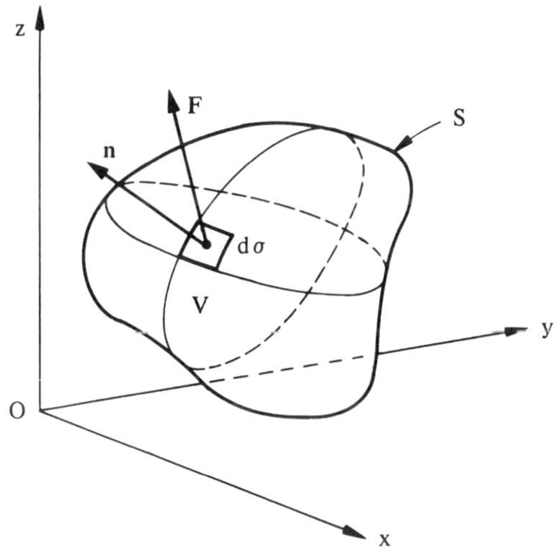

Fig. 10.1

As an example, let V be the sphere of radius R centered at the origin and $\mathbf{F} = \mathbf{r} = x\mathbf{i} + y\mathbf{j} + z\mathbf{k}$. According to the divergence theorem (10.1), we have

$$\iiint\limits_V \operatorname{div} \mathbf{F}\, dV = 3\iiint\limits_V dV = 3V = \oiint\limits_S \mathbf{r}\cdot\mathbf{n}\, d\sigma$$

and since we have $\mathbf{r}\cdot\mathbf{n} = R$ on the surface of the sphere of radius R, we get

$$3V = R \oiint\limits_S d\sigma = 4\pi R^3$$

so that for the volume of the sphere, we obtain

$$V = \frac{4}{3}\pi R^3$$

Before proving formula (10.1), we shall proceed to the following example.

10.1.1 Example: Verification of the divergence theorem

Let us verify the divergence theorem for the case in which $\mathbf{F} = xyz\mathbf{k}$ and V is the tetrahedron bounded by the four planes $x = 0$, $y = 0$, $z = 0$, and $x + y + z = 1$ (fig. 10.2).

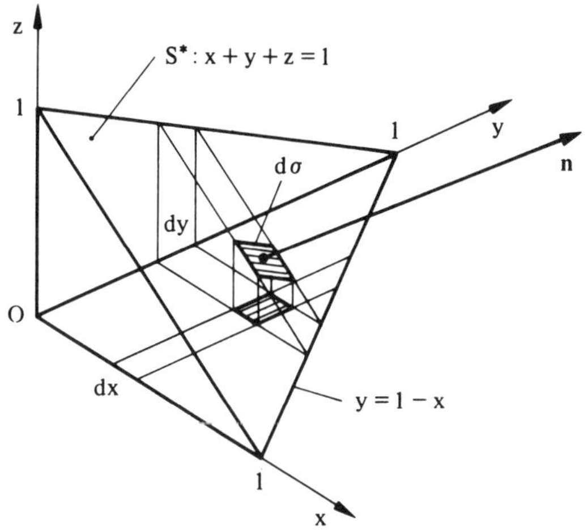

Fig. 10.2

The surface integral must be taken over the four triangles forming the boundaries of the tetrahedron. Because $\mathbf{F} = \mathbf{0}$ on the surfaces $x = 0$, $y = 0$, $z = 0$, only the surface integral over the triangular piece S*: $x + y + z = 1$, $0 \leq y \leq 1 - x$, $0 \leq x \leq 1$ remains. Choosing x and y as parameters, S* has the following vector representation:

$$\mathbf{r}(x, y) = x\mathbf{i} + y\mathbf{j} + (1 - x - y)\mathbf{k}$$

and, according to formula (10.5), we find for $\mathbf{n}d\sigma$ (such that $\mathbf{n}$ is directed toward the exterior of the tetrahedron):

$$\mathbf{n}\, d\sigma = \left(\frac{\partial \mathbf{r}}{\partial x} \wedge \frac{\partial \mathbf{r}}{\partial y} \right) dx\, dy = (\mathbf{i} + \mathbf{j} + \mathbf{k})\, dx\, dy$$

Thus, we obtain for the surface integral:

$$\iint_{S^*} \mathbf{F} \cdot \mathbf{n}\, d\sigma = \int_{x=0}^{1} \left\{ \int_{y=0}^{1-x} xyz\, dy \right\} dx$$

$$= \int_{x=0}^{1} \left\{ \int_{y=0}^{1-x} xy\,(1 - x - y)\, dy \right\} dx = \frac{1}{120}$$

For the volume integral, we obtain

$$\iiint_V \operatorname{div} \mathbf{F}\, dV = \iiint_V xy\, dV = \int_{x=0}^{1} \left\{ \int_{y=0}^{1-x} \left\{ \int_{z=0}^{1-x-y} xy\, dz \right\} dy \right\} dx = \frac{1}{120}$$

Thus, we have verified the divergence theorem for this particular situation.

10.2 DEMONSTRATION OF THE DIVERGENCE THEOREM

We demonstrate the divergence theorem for the case in which S is a closed surface such that any line parallel to the x,y, or z coordinate axes cuts S in at most two points (fig. 10.3).

S is divided into an upper part S_1 given by the equation:

$$S_1 : z = z_1(x, y)$$

and a lower part S_0 satisfying the equation:

$$S_0 : z = z_0(x, y)$$

Let $\mathbf{n}$ be the unit vector normal to the surface, directed toward the exterior of S. On the upper part S_1 of S, the normal vector $\mathbf{n}_1$ forms an acute angle γ_1 with the z-axis, and on the lower part S_0, the vector $\mathbf{n}_0$ forms an acute angle γ_0 with the negative z-axis (fig. 10.3).

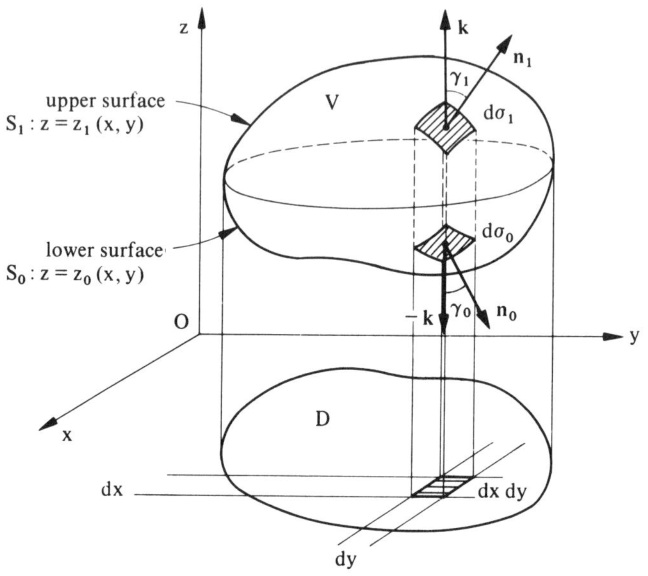

Fig. 10.3

Let $\mathbf{F}$ be a vector field defined on V and S, having continuous partial derivatives:

$$\mathbf{F}(x, y, z) = F_x(x, y, z)\mathbf{i} + F_y(x, y, z)\mathbf{j} + F_z(x, y, z)\mathbf{k}$$

From definition (7.23), we have

$$\operatorname{div} \mathbf{F} = \frac{\partial F_x}{\partial x} + \frac{\partial F_y}{\partial y} + \frac{\partial F_z}{\partial z} \tag{10.2}$$

Let us first consider the triple integral over the volume V of the third term on the right-hand side of (10.2):

$$\iiint_V \frac{\partial F_7}{\partial z}\, dV = \iint_D \left\{ \int_{z_0}^{z_1} \frac{\partial F_z}{\partial z}\, dz \right\} dx\, dy \tag{10.3}$$

where the double integral with respect to x and y is taken over the orthogonal projection of the surface S on the xy-plane. On the upper part S_1 of S, we have

$$\cos \gamma_1\, d\sigma_1 = \mathbf{k} \cdot \mathbf{n}_1\, d\sigma_1 = dx\, dy \tag{10.4}$$

and on the lower part S_0:

$$\cos\gamma_0\, d\sigma_0 = -\mathbf{k}\cdot\mathbf{n}_0\, d\sigma_0 = dx\, dy \tag{10.5}$$

since the normal $\mathbf{n}_0$ forms an obtuse angle with $\mathbf{k}$. Thus, using relations (10.4) and (10.5), we obtain for the volume integral:

$$\iiint_V \frac{\partial F_z}{\partial z}\, dV = \iint_D [F_z(x, y, z_1) - F_z(x, y, z_0)]\, dx\, dy$$

$$= \iint_{S_1} F_z(x, y, z)\mathbf{k}\cdot\mathbf{n}\, d\sigma + \iint_{S_0} F_z(x, y, z)\mathbf{k}\cdot\mathbf{n}\, d\sigma \tag{10.6}$$

It should be noted that the indices of the vector $\mathbf{n}$ are omitted in the integrals on the right-hand side of equation (10.6), since the parts of the surface over which integration is carried out are designated by the domain of integration under the integral sign. The right side of (10.6) is the sum of an integral over the lower part and of an integral over the upper part of the surface S, i.e., we can write

$$\iiint_V \frac{\partial F_z}{\partial z}\, dV = \oint\!\!\oint_S F_z\, \mathbf{k}\cdot\mathbf{n}\, d\sigma \tag{10.7}$$

Likewise, by orthogonal projection of S on the two other coordinate planes, we arrive at

$$\iiint_V \frac{\partial F_x}{\partial x}\, dV = \oint\!\!\oint_S F_x\, \mathbf{i}\cdot\mathbf{n}\, d\sigma \tag{10.8}$$

and

$$\iiint_V \frac{\partial F_y}{\partial y}\, dV = \oint\!\!\oint_S F_y\, \mathbf{j}\cdot\mathbf{n}\, d\sigma \tag{10.9}$$

Now, by forming the sum of relations (10.7), (10.8), and (10.9), we arrive at the formula for the divergence theorem (10.1):

$$\iiint_V \left(\frac{\partial F_x}{\partial x} + \frac{\partial F_y}{\partial y} + \frac{\partial F_z}{\partial z}\right) dV = \iiint_V \operatorname{div}\mathbf{F}\, dV$$

$$= \oint\!\!\oint_S (F_x\,\mathbf{i} + F_y\,\mathbf{j} + F_z\,\mathbf{k})\cdot\mathbf{n}\, d\sigma = \oint\!\!\oint_S \mathbf{F}\cdot\mathbf{n}\, d\sigma$$

10.3 GRADIENT THEOREM

The gradient theorem is a corollary of the divergence theorem. If we set $\mathbf{F} = \phi\mathbf{a}$ in formula (10.1), where $\phi = \phi\,(x,y,z)$ is a scalar field defined on V and S, having continuous partial derivatives, and **a** is an arbitrary constant vector, then

$$\iiint_V \operatorname{div}(\phi\, \mathbf{a})\, dV = \oiint_S (\phi\, \mathbf{a}) \cdot \mathbf{n}\, d\sigma = \mathbf{a} \cdot \oiint_S \phi\, \mathbf{n}\, d\sigma \tag{10.10}$$

Because (7.33) states

$$\operatorname{div}(\phi\, \mathbf{a}) = \mathbf{a} \cdot \mathbf{grad}\, \phi$$

we obtain for the left side of relation (10.10), moving the constant vector **a** across the integral sign:

$$\mathbf{a} \cdot \iiint_V \mathbf{grad}\, \phi\, dV = \mathbf{a} \cdot \oiint_S \phi\, \mathbf{n}\, d\sigma$$

Since **a** is an arbitrary vector, it follows that

$$\iiint_V \mathbf{grad}\, \phi\, dV = \oiint_S \phi\, \mathbf{n}\, d\sigma \tag{10.11}$$

which constitutes the ***gradient theorem***.

10.4 PHYSICAL APPLICATIONS

10.4.1 Archimedes' principle

Archimedes' principle states that any body immersed in a fluid undergoes a vertical force equal to the weight of the displaced fluid.

The demonstration of this principle consists of calculating the resultant of the forces of pressure against the entire surface S of the immersed body of volume V.

The differential force of pressure **dF** acting on a differential surface element $d\sigma$ is:

$$d\mathbf{F} = -p\, \mathbf{n}\, d\sigma \tag{10.12}$$

where p designates the pressure. Relation (10.12) is valid on the condition that there are no tangential forces along the surface of the body, which assumes that the viscosity of the fluid is zero. The force **F** acting over the entire surface S is the closed surface integral:

$$\mathbf{F} = \oiint_S d\mathbf{F} = -\oiint_S p\, \mathbf{n}\, d\sigma$$

and applying the gradient theorem (10.11), we have

$$\mathbf{F} = -\oiint_S p\,\mathbf{n}\,d\sigma = -\iiint_V \mathbf{grad}\,p\,dV$$

Assuming that the specific weight γ of the fluid is constant (i.e., it is an incompressible fluid), we have for the pressure p:

$$p = p_0 - \gamma z$$

where p_0 designates the pressure at the surface $z = 0$ of the fluid and $z < 0$ is the depth of the volume element dV. It follows that

$$\mathbf{grad}\,p = -\gamma\mathbf{k}$$

and, consequently,

$$\mathbf{F} = \mathbf{k}\iiint_V \gamma\,dV \tag{10.13}$$

which expresses ***Archimedes' principle***.

10.4.2 Equation of continuity

Let $\mathbf{v}$ be the velocity vector of a fluid of density ρ that crosses the closed permeable surface S which bounds the domain V in space. At any instant t, the mass of fluid M inside V is given by the volume integral:

$$M(t) = \iiint_V \rho(x, y, z, t)\,dV$$

The rate of variation of the mass M equals

$$\frac{dM}{dt} = \iiint_V \frac{\partial\rho}{\partial t}\,dV$$

Let us assume that there is an increase in the mass, i.e., that $dM/dt > 0$, and that S is a sphere (fig. 10.4).

The mass of liquid entering the sphere, per unit time, is given by the surface integral over S_1 of the velocity vector:

$$-\iint_{S_1} \rho\,\mathbf{v}\cdot\mathbf{n}\,d\sigma$$

and that of the fluid exiting the sphere is determined by:

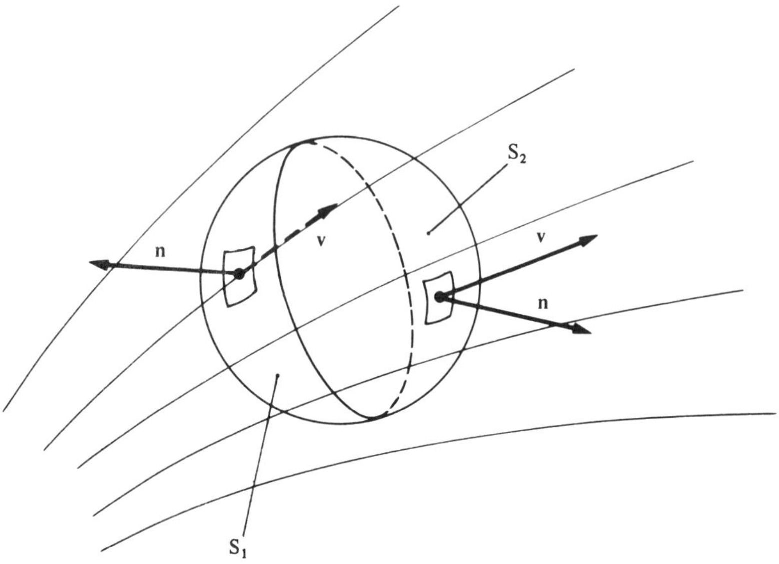

Fig. 10.4

$$\iint_{S_2} \rho\, \mathbf{v} \cdot \mathbf{n}\, d\sigma$$

Thus, the balance is as follows:

$$\frac{dM}{dt} = -\iint_{S_1} \rho\, \mathbf{v} \cdot \mathbf{n}\, d\sigma - \iint_{S_2} \rho\, \mathbf{v} \cdot \mathbf{n}\, d\sigma = -\oint\!\!\!\oint_{S} \rho\, \mathbf{v} \cdot \mathbf{n}\, d\sigma = \iiint_{V} \frac{\partial \rho}{\partial t}\, dV$$

From the divergence theorem (10.1), we have

$$-\oint\!\!\!\oint_{S} \rho\, \mathbf{v} \cdot \mathbf{n}\, d\sigma = -\iiint_{V} \operatorname{div}(\rho\, \mathbf{v})\, dV = \iiint_{V} \frac{\partial \rho}{\partial t}\, dV$$

from which we get

$$\iiint_{V} \left(\operatorname{div}(\rho\, \mathbf{v}) + \frac{\partial \rho}{\partial t} \right) dV = 0 \qquad (10.14)$$

Because the considerations remain the same if there is a decrease in the mass, i.e., $dM/dt < 0$, relation (10.14) is valid for any variation of mass. Further, (10.14) is valid for any sphere at the interior of the liquid, which means, if it is assumed that the integrated functions in the integral (10.14) are continuous, that the integrand must vanish identically. Thus, we find the following relation, which relates the density and the velocity vector of a liquid:

$$\operatorname{div}(\rho\,\mathbf{v}) + \frac{\partial \rho}{\partial t} = 0 \tag{10.15}$$

Equation (10.15) is called the ***equation of continuity***, which expresses the condition for the conservation of mass. If ρ is constant, the fluid is incompressible and the equation of continuity is reduced to the condition of ***incompressibility***:

$$\operatorname{div}\mathbf{v} = 0 \tag{10.16}$$

10.4.3 Example: Moment of inertia of a sphere

The moment of inertia I_0 of a homogeneous sphere, of density ρ and radius R, with respect to its center is given by the volume integral:

$$I_0 = \rho \iiint_V r^2\, dV \tag{10.17}$$

Let us consider the vector function:

$$\mathbf{F} = \frac{r^2\,\mathbf{r}}{5}$$

where $\mathbf{r}$ designates the radius vector of a point P and r is its magnitude. For the divergence of $\mathbf{F}$, we have

$$\operatorname{div}\mathbf{F} = \frac{1}{5}\operatorname{div}[(x^2 + y^2 + z^2)(x\mathbf{i} + y\mathbf{j} + z\mathbf{k})]$$

$$= \frac{1}{5}(3x^2 + y^2 + z^2 + x^2 + 3y^2 + z^2 + x^2 + y^2 + 3z^2)$$

$$= x^2 + y^2 + z^2 = r^2$$

Applying the divergence theorem (10.1) to the volume integral (10.17), we find

$$I_0 = \rho \iiint_V r^2\, dV - \rho \iiint_V \operatorname{div}\left(\frac{r^2\,\mathbf{r}}{5}\right) dV -$$

$$= \rho \oiint_S \frac{r^2}{5}\,\mathbf{r}\cdot\mathbf{n}\, d\sigma = \frac{\rho R^3}{5} \oiint_S d\sigma = \frac{4 R^5 \pi \rho}{5} = \frac{3}{5} MR^2$$

where M designates the total mass of the sphere.

10.5 GREEN'S THEOREM IN SPACE

Let ϕ and ψ be two continuous scalar fields having continuous second-order partial derivatives in a domain V and on its boundary S. We shall apply the divergence theorem (10.1) to the vector function **F**:

$$\mathbf{F} = \psi\, \mathbf{grad}\, \phi - \phi\, \mathbf{grad}\, \psi$$

As (7.33) states:

$$\operatorname{div} \mathbf{F} = \mathbf{grad}\, \psi \cdot \mathbf{grad}\, \phi + \psi \operatorname{div} \mathbf{grad}\, \phi$$
$$- \mathbf{grad}\, \phi \cdot \mathbf{grad}\, \psi - \phi \operatorname{div} \mathbf{grad}\, \psi = \psi \triangle \phi - \phi \triangle \psi$$

we find

$$\iiint_V \operatorname{div} \mathbf{F}\, dV = \iiint_V (\psi \triangle \phi - \phi \triangle \psi)\, dV$$

$$= \oiint_S (\psi\, \mathbf{grad}\, \phi - \phi\, \mathbf{grad}\, \psi) \cdot \mathbf{n}\, d\sigma \qquad (10.18)$$

The scalar products **grad** $\phi \cdot$ **n** and **grad** $\psi \cdot$ **n** in the integral on the right side of equation (10.18) are the derivatives of the functions ϕ and ψ, respectively, in the direction **n**, written $\partial\phi/\partial n$ and $\partial\psi/\partial n$. From (10.18), we then have

$$\iiint_V (\psi \triangle \phi - \phi \triangle \psi)\, dV = \oiint_S \left(\psi \frac{\partial \phi}{\partial n} - \phi \frac{\partial \psi}{\partial n} \right) d\sigma \qquad (10.19)$$

which constitutes ***Green's theorem in space.***

Chapter 11

Stokes' Theorem and Applications

11.1 STOKES' THEOREM

Consider a two-sided surface S having a simple closed curve C as its boundary and a vector field $\mathbf{F}(x, y, z)$ defined on S and C, having continuous partial derivatives (fig. 11.1). Then, ***Stokes' theorem*** states that the closed line integral of **F** along the contour C taken in the positive direction is equal to the surface integral of the curl of **F** over S:

$$\oint_C \mathbf{F} \cdot d\mathbf{r} = \iint_S \mathbf{curl}\ \mathbf{F} \cdot \mathbf{n}\, d\sigma \tag{11.1}$$

The direction or sense of C is called positive, if an observer traveling along C, with his head pointing in the direction of the vector **n** normal to the surface, has the surface on his left.

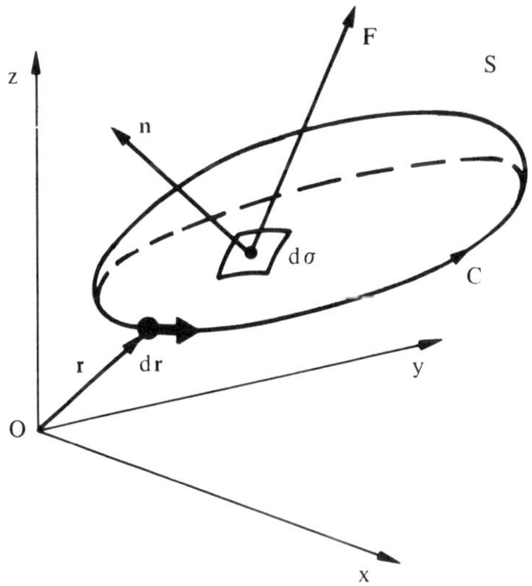

Fig. 11.1

According to Stokes' theorem, the surface integral:

$$\iint_S \mathbf{curl\ F} \cdot \mathbf{n}\, d\sigma$$

depends only on **F** along the contour C of S.

Before demonstrating formula (11.1), we shall proceed with the following example.

11.1.1 Example: Verification of Stokes' theorem

Let $\mathbf{F} = y\mathbf{i} + 2x\mathbf{j} + z\mathbf{k}$, and C be the circle given by $x^2 + y^2 = 1$, $z = 0$. We consider C to be either the boundary of the paraboloid $z = 1 - x^2 - y^2$, $z \geq 0$, or the boundary of the hemisphere $x^2 + y^2 + z^2 = 1$, $z \geq 0$. The parametric equation of the circle C is

$$\mathbf{r}(\varphi) = \cos\varphi\mathbf{i} + \sin\varphi\mathbf{j}, \quad 0 \leq \varphi \leq 2\pi$$

we calculate for the line integral:

$$\oint_C \mathbf{F} \cdot d\mathbf{r} = \int_0^{2\pi} (\sin\varphi\mathbf{i} + 2\cos\varphi\mathbf{j})\,(-\sin\varphi\mathbf{i} + \cos\varphi\mathbf{j})\, d\varphi$$

$$= \int_0^{2\pi} (-\sin^2\varphi + 2\cos^2\varphi)\, d\varphi = \pi$$

Because we have

$$\mathbf{curl\ F} = \mathbf{k}$$

the surface integral becomes

$$\iint_S \mathbf{curl\ F} \cdot \mathbf{n}\, d\sigma = \iint_S \mathbf{k} \cdot \mathbf{n}\, d\sigma$$

If S designates the hemisphere of radius 1, we have (section 9.2):

$$\mathbf{n}\, d\sigma = \mathbf{r} \sin\vartheta\, d\vartheta\, d\varphi$$

and taking into account

$$\mathbf{r} \cdot \mathbf{k} = \cos\vartheta$$

we find for the integral of **curl F** over the hemisphere:

$$\iint_S \mathbf{curl\ F} \cdot \mathbf{n}\, d\sigma = \int_{\varphi=0}^{2\pi} \int_{\vartheta=0}^{\pi/2} \sin\vartheta \cos\vartheta\, d\vartheta\, d\varphi = \pi$$

To evaluate the integral over the paraboloid, we represent the surface by means of cylindrical coordinates:

$$\mathbf{r}(\rho, \varphi) = \rho \cos \varphi \mathbf{i} + \rho \sin \varphi \mathbf{j} + (1 - \rho^2)\,\mathbf{k}$$

where the domain of variation of the parameters φ and ρ is determined by the inequalities $0 \leq \rho \leq 1$ and $0 \leq \varphi \leq 2\pi$. We calculate

$$\mathbf{n}\, d\sigma = \left(\frac{\partial \mathbf{r}}{\partial \rho} \wedge \frac{\partial \mathbf{r}}{\partial \varphi} \right) d\rho\, d\varphi = (2\rho^2 \cos \varphi \mathbf{i} + 2\rho^2 \sin \varphi \mathbf{j} + \rho \mathbf{k})\, d\rho\, d\varphi$$

and obtain

$$\iint_S \mathbf{curl}\ \mathbf{F} \cdot \mathbf{n}\, d\sigma = \int_{\varphi=0}^{2\pi} \int_{\rho=0}^{1} \rho\, d\rho\, d\varphi = \pi$$

Thus, Stokes' theorem has been verified for this particular situation.

11.2 GREEN'S THEOREM IN THE PLANE

With the aim of proving Stokes' theorem, we first state ***Green's theorem in the plane***, which expresses the following fact. Let D be a simply connected region in the xy-plane bounded by the simple closed curve C, and let P (x, y) and Q (x, y) be two continous functions having continuous partial derivatives in D and on C. Then,

$$\oint_C (P\, dx + Q\, dy) = \iint_D \left(\frac{\partial Q}{\partial x} - \frac{\partial P}{\partial y} \right) dx\, dy \qquad (11.2)$$

where the integration along C proceeds in the positive direction, i.e., such that the domain D is found to the left of an observer traveling along C.

We shall demonstrate formula (11.2) for a closed curve C, which has the property that any straight line parallel to the coordinate axes cuts C in at most two points (fig. 11.2).

Let $y = y_1(x)$ be the equation of the lower arc AEB and $y = y_2(x)$ that of the upper arc AFB. For the double integral of $\partial P/\partial y$ over D, we have

$$\iint_D \frac{\partial P}{\partial y}\, dx\, dy = \int_a^b \left\{ \int_{y_1}^{y_2} \frac{\partial P}{\partial y}\, dy \right\} dx$$

$$= \int_a^b \{P(x, y_2) - P(x, y_1)\}\, dx$$

$$\iint_D \frac{\partial P}{\partial x}\, dx\, dy = -\left[\int_b^a P(x, y_2)\, dx + \int_a^b P(x, y_1)\, dx\right]$$

$$= -\oint_C P(x, y)\, dx \qquad (11.3)$$

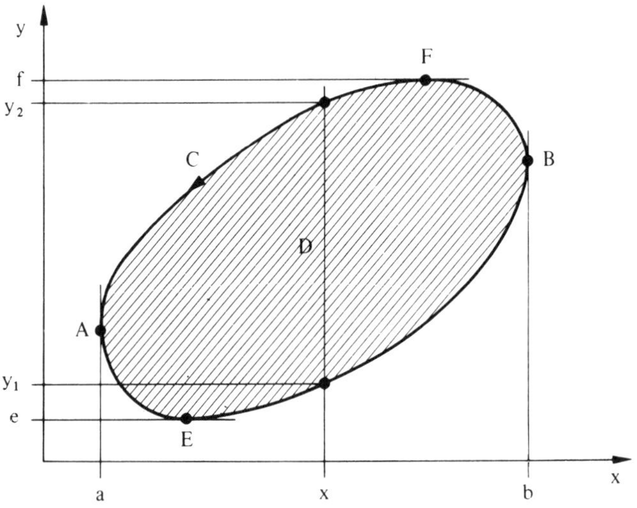

Fig. 11.2

In an analogous fashion, if the equations of the arcs EAF and EBF are given as a function of y, we find for the double integral of $\partial Q / \partial x$ over D:

$$\iint_D \frac{\partial Q}{\partial x}\, dx\, dy = \oint_C Q\, dy \qquad (11.4)$$

By subtracting equation (11.3) from (11.4), we obtain Green's theorem in the plane (11.2).

11.3 DEMONSTRATION OF STOKES' THEOREM

To simplify the demonstration, we will assume that the surface S is such that any line parallel to a coordinate axis cuts it at a single point (fig. 11.3).

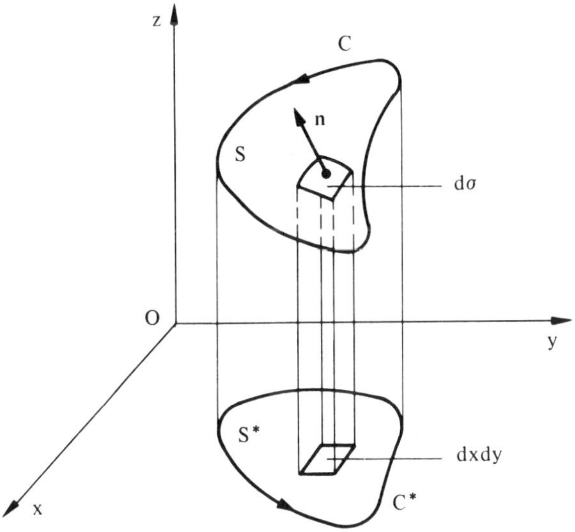

Fig. 11.3

C designates the boundary of S traversed in the positive direction. Let us assume that S is represented by $z = z\,(x, y)$, where the function z is single-valued and has continuous partial derivatives; the domain of variation of the parameters x and y is the orthogonal projection S* of the surface S on the xy-plane. For the position vector **r** of a point P of S, we have

$$\mathbf{r} = x\mathbf{i} + y\mathbf{j} + z(x, y)\,\mathbf{k}$$

and, consequently,

$$\frac{\partial \mathbf{r}}{\partial x} = \mathbf{i} + \frac{\partial z}{\partial x}\,\mathbf{k}$$

$$\frac{\partial \mathbf{r}}{\partial y} = \mathbf{j} + \frac{\partial z}{\partial y}\,\mathbf{k}$$

such that

$$\mathbf{n}\,d\sigma = \left(\frac{\partial \mathbf{r}}{\partial x} \wedge \frac{\partial \mathbf{r}}{\partial y}\right) dx\,dy = \left(-\frac{\partial z}{\partial x}\,\mathbf{i} - \frac{\partial z}{\partial y}\,\mathbf{j} + \mathbf{k}\right) dx\,dy \tag{11.5}$$

Let us now consider the vector field:

$$\mathbf{F} = F_x\,\mathbf{i} + F_y\,\mathbf{j} + F_z\,\mathbf{k}$$

defined on S and C, and having continuous derivatives. We have

$$\mathbf{curl}\ \mathbf{F} = \mathbf{curl}\,(F_x\,\mathbf{i}) + \mathbf{curl}\,(F_y\,\mathbf{j}) + \mathbf{curl}\,(F_z\,\mathbf{k}) \tag{11.6}$$

The first term on the right side of equation (11.6) equals

$$\mathbf{curl}\,(F_x\,\mathbf{i}) = \frac{\partial F_x}{\partial z}\,\mathbf{j} - \frac{\partial F_x}{\partial y}\,\mathbf{k}$$

and using equation (11.5), we find

$$\mathbf{curl}\,(F_x\,\mathbf{i}) \cdot \mathbf{n}\, d\sigma = -\left(\frac{\partial F_x}{\partial z}\,\frac{\partial z}{\partial y} + \frac{\partial F_x}{\partial y}\right) dx\, dy \tag{11.7}$$

Then, on the surface S, we set

$$F_x\,(x, y, z) = F_x\,(x, y, z(x, y)) = P\,(x, y) \tag{11.8}$$

from which we get

$$\frac{\partial P}{\partial y} = \frac{\partial F_x}{\partial y} + \frac{\partial F_x}{\partial z}\,\frac{\partial z}{\partial y} \tag{11.9}$$

and it follows that (11.7) takes the form:

$$\mathbf{curl}\,(F_x\,\mathbf{i}) \cdot \mathbf{n}\, d\sigma = -\frac{\partial P}{\partial y}\, dx\, dy$$

Thus, we obtain for the surface integral of the curl:

$$\iint_S \mathbf{curl}\,(F_x\,\mathbf{i}) \cdot \mathbf{n}\, d\sigma = -\iint_{S^*} \frac{\partial P}{\partial y}\, dx\, dy \tag{11.10}$$

Applying Green's formula (11.2) to the right side of equation (11.10), we find

$$\iint_S \mathbf{curl}\,(F_x\,\mathbf{i}) \cdot \mathbf{n}\, d\sigma = \oint_{C^*} P\, dx$$

where C* is the contour of S* in the xy-plane. According to definition (11.8) of the function $P(x, y)$, the value of $P(x, y)$ at any point (x, y) of the contour C* of S* is the value of $F_x(x, y, z)$ at the point (x, y, z) of the boundary C of S. Thus, we arrive at

$$\iint_S \mathbf{curl}\,(F_x\,\mathbf{i}) \cdot \mathbf{n}\, d\sigma = \oint_C F_x\, dx \tag{11.11}$$

Likewise, by orthogonal projection of S on the two other coordinate planes, we obtain

$$\iint_S \mathbf{curl}\,(F_y\,\mathbf{j}) \cdot \mathbf{n}\, d\sigma = \oint_C F_y\, dy \tag{11.12}$$

and

$$\iint_S \mathbf{curl}\,(F_z\,\mathbf{k}) \cdot \mathbf{n}\, d\sigma = \oint_C F_z\, dz \tag{11.13}$$

such that considering the sum of relations (11.11), (11.12), and (11.13), we arrive at Stokes' theorem:

$$\iint_S \mathbf{curl}\,\mathbf{F} \cdot \mathbf{n}\, d\sigma = \oint_C (F_x\, dx + F_y\, dy + F_z\, dz) = \oint_C \mathbf{F} \cdot d\mathbf{r}$$

11.4 EXISTENCE OF A POTENTIAL OF A VECTOR FIELD

Let $\mathbf{F} = \mathbf{F}(x, y, z)$ be a vector field possessing continuous derivatives. The condition that the line integral of **F** between two points is independent of the path joining those two points is equivalent to the condition that the line integral of **F** along any closed curve C vanishes:

$$\oint_C \mathbf{F} \cdot d\mathbf{r} = 0 \tag{11.14}$$

Using Stokes' formula (11.1), equation (11,14) takes the form:

$$\oint_C \mathbf{F} \cdot d\mathbf{r} = \iint_S \mathbf{curl}\,\mathbf{F} \cdot \mathbf{n}\, d\sigma = 0 \tag{11.15}$$

where S is any surface bounded by the closed curve C. Thus,

$$\mathbf{curl}\,\mathbf{F} = \mathbf{0} \tag{11.16}$$

is a sufficient condition for the line integral of **F** along any closed curve to cancel out.

Conversely, if condition (11.14) is satisfied, we can construct a scalar function ϕ such that

$$\mathbf{F} = \mathbf{grad}\,\phi$$

i.e., the function ϕ is the ***potential*** of the vector field **F**. The potential ϕ is constructed in the following manner:

$$\phi\,(x, y, z) = \int_{(x_0, y_0, z_0)}^{(x, y, z)} \mathbf{F} \cdot d\mathbf{r} \tag{11.17}$$

where the initial point (x_0, y_0, z_0) is arbitrarily chosen and the integral is taken along any curve linking the points (x_0, y_0, z_0) and (x, y, z). Let us choose a broken line as a path, composed of three segments: a segment parallel to the z-axis joining the point (x_0, y_0, z_0) to the point (x_0, y_0, z), a segment parallel to the y-axis leading from the point (x_0, y_0, z) to the point (x_0, y, z), and a segment parallel to the x-axis linking the two points (x_0, y, z) and (x, y, z) (fig. 11.4). Thus, we find for the line integral (11.17):

$$\phi\,(x, y, z) = \int_{x_0}^{x} F_x\,(\xi, y, z)\,d\xi + \int_{y_0}^{y} F_y\,(x_0, \tau, z)\,d\tau + \int_{z_0}^{z} F_z\,(x_0, y_0, \zeta)\,d\zeta \tag{11.18}$$

We can immediately see that

$$\frac{\partial \phi}{\partial x} = F_x$$

since the second and third integrals on the right side of (11.18) do not depend on x. Changing the segments of the broken line, we verify that

$$\frac{\partial \phi}{\partial y} = F_y$$

and

$$\frac{\partial \phi}{\partial z} = F_z$$

with the result that we have, in vector notation:

$$\mathbf{F} = \mathbf{grad}\ \phi$$

We, therefore, have (section 7.4):

$$\mathbf{curl\ F} = \mathbf{curl\ grad}\ \phi = \mathbf{0}$$

i.e., condition (11.16):

$$\mathbf{curl\ F} = \mathbf{0}$$

which is also a necessary condition for the line integral of **F** along any closed curve to vanish.

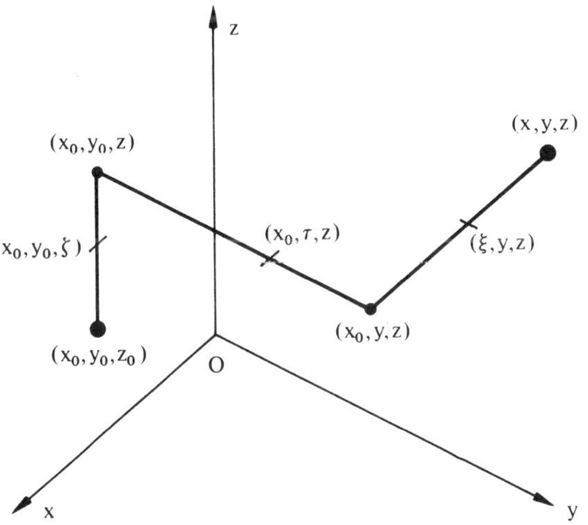

Fig. 11.4

We can immediately see that the line integral of $\mathbf{F} = F_x\mathbf{i} + F_y\mathbf{j} + F_z\mathbf{k}$ is independent of the path of integration if and only if the differential form:

$$F_x\,dx + F_y\,dy + F_z\,dz$$

is the total differential $d\phi$ of a function ϕ.

In other words, the line integral of **F** is independent of the path of integration if and only if **F** has a potential ϕ.

Thus, we have just seen that the following ***propositions*** are ***equivalent***:

- the integral $\int_{P_0}^{P_1} \mathbf{F}\cdot d\mathbf{r}$ is independent on the path linking P_0 and P_1;
- the integral $\int_C \mathbf{F}\cdot d\mathbf{r}$ vanishes along any closed curve C;
- **curl F = 0**;
- $F_x dx + F_y dy + F_z dz$ is the total differential $d\phi$ of a function ϕ.
- **F** has a potential ϕ.

11.4.1 Example: Independence of a line integral of the path of integration

Consider the vector field:

$$\mathbf{F} = -\frac{y}{x^2 + y^2}\,\mathbf{i} + \frac{x}{x^2 + y^2}\,\mathbf{j}$$

whose components and their derivatives are continuous in the whole space except at the points where $x^2 + y^2 = 0$, i.e., on the z-axis. We have

$$\mathbf{curl\ F} = \begin{vmatrix} \mathbf{i} & \mathbf{j} & \mathbf{k} \\ \dfrac{\partial}{\partial x} & \dfrac{\partial}{\partial y} & \dfrac{\partial}{\partial z} \\ \dfrac{-y}{x^2+y^2} & \dfrac{x}{x^2+y^2} & 0 \end{vmatrix} = \left\{ \frac{\partial}{\partial x}\left(\frac{x}{x^2+y^2}\right) + \frac{\partial}{\partial y}\left(\frac{y}{x^2+y^2}\right) \right\} \mathbf{k}$$

and for $x^2 + y^2 > 0$, we verify that **curl F** = **0**.

Let us consider the line integral of **F** along the arbitrary closed curve C:

$$\oint_C \mathbf{F} \cdot d\mathbf{r} = \oint_C \frac{x\,dy - y\,dx}{x^2 + y^2} = \oint_C d\left(\text{arc tan}\,\frac{y}{x}\right) = \varphi \tag{11.19}$$

This integral represents the angle φ through which the orthogonal projection of the position vector of a point P of C in the xy-plane turns as P traverses C. If a closed curve C encircles the z-axis, the angle φ is a multiple of 2π; otherwise, φ vanishes. All of this is in perfect agreement with Stokes' theorem (11.1):

$$\oint_C \mathbf{F} \cdot d\mathbf{r} = \iint_S \mathbf{curl\ F} \cdot \mathbf{n}\, d\sigma$$

If C encircles the z-axis, any surface with the boundary C is cut by the z-axis, and this implies that there are points on S at which **curl F** is not defined. Consequently, integral (11.19) does not necessarily vanish. When the curve C does not encircle the z-axis, on all the surfaces S of boundary C we have **curl F** = **0**, and integral (11.19) vanishes accordingly.

11.4.2 Example: Total differential

Consider the differential form:

$$(y^2 z^3 \cos x - 4x^3 z)\, dx + 2yz^3 \sin x\, dy + (3y^2 z^2 \sin x - x^4)\, dz$$

We set

$$\mathbf{F} = (y^2 z^3 \cos x - 4x^3 z)\,\mathbf{i} + 2yz^3 \sin x\,\mathbf{j} + (3y^2 z^2 \sin x - x^4)\,\mathbf{k}$$

and verify that **curl F** = **0**. Thus, the differential form given is the total differential of a potential ϕ. We know that

$$\frac{\partial \phi}{\partial x} = y^2 z^3 \cos x - 4x^3 z \tag{11.20}$$

$$\frac{\partial \phi}{\partial y} = 2yz^3 \sin x \tag{11.21}$$

$$\frac{\partial \phi}{\partial z} = 3y^2 z^2 \sin x - x^4 \tag{11.22}$$

and, in order to determine the function ϕ, we proceed in the following manner. The first equation (11.20) is integrated with respect to the variable x, and we obtain

$$\phi = y^2 z^3 \sin x - x^4 z + g\,(y, z) \tag{11.23}$$

where the function g (y,z) is the integration "constant." Then we differentiate this equation (11.23) with respect to y, and equating the result with (11.21), we find

$$2yz^3 \sin x + \frac{\partial g}{\partial y} = 2yz^3 \sin x$$

from which we get

$$\frac{\partial g}{\partial y} = 0$$

It follows that

$$g = h\,(z)$$

and substituting this result into equation (11.23), we have

$$\phi = y^2 z^3 \sin x - x^4 z + h\,(z) \tag{11.24}$$

Finally, we differentiate this equation (11.24) with respect to z and compare the result with equation (11.22):

$$\frac{\partial \phi}{\partial z} = 3y^2 z^2 \sin x - x^4 + \frac{dh}{dz} = 3y^2 z^2 \sin x - x^4$$

from which we get

$$\frac{dh}{dz} = 0$$

i.e., h is a constant. Thus, we find the potential ϕ:

$$\phi = y^2 z^3 \sin x - x^4 z + \text{constant}$$

We may easily verify that

$$d\phi = (y^2 z^3 \cos x - 4x^3 z)\, dx + 2yz^3 \sin x\, dy + (3y^2 z^3 \sin x - x^4)\, dz$$

Chapter 12

Orthogonal Curvilinear Coordinates

12.1 CURVILINEAR COORDINATES

Until now we have only used rectangular Cartesian coordinates x, y, and z. Let us assume now that the rectangular Cartesian coordinates x, y, and z can be expressed in terms of new coordinates u, v, and w by the equations:

$$\begin{aligned} x &= x(u, v, w) \\ y &= y(u, v, w) \\ z &= z(u, v, w) \end{aligned} \tag{12.1}$$

Let us further assume that these relations can be inverted to express u, v, w in terms of x, y, z. Then, to any point with coordinates (x, y, z), there corresponds a point with coordinates (u, v, w):

$$\begin{aligned} u &= u(x, y, z) \\ v &= v(x, y, z) \\ w &= w(x, y, z) \end{aligned} \tag{12.2}$$

We assume that the correspondence (12.1) and (12.2) is unique.

If a particle moves from the point P (u_0, v_0, w_0) in such a way that v and w are held constant while u varies, a curve is generated in the corresponding xyz coordinate system. We refer to this curve as the u curve. Similarly, two other curves, the v and w curves, are generated at each point P (fig. 12.1).

It is often convenient to choose the new coordinates u, v, w in such a way that the coordinate curves are mutually perpendicular at each point in space. Such coordinates are called ***orthogonal curvilinear coordinates***.

Let $\mathbf{r} = x\mathbf{i} + y\mathbf{j} + z\mathbf{k}$ be the position vector of a point P (x, y, z). Given equations (12.1), we can also consider this vector as a function of u, v, and w: $\mathbf{r} = \mathbf{r}(u, v, w)$. Then, the vector:

$$\frac{\partial \mathbf{r}}{\partial u} = \frac{\partial x}{\partial u}\mathbf{i} + \frac{\partial y}{\partial u}\mathbf{j} + \frac{\partial z}{\partial u}\mathbf{k} \tag{12.3}$$

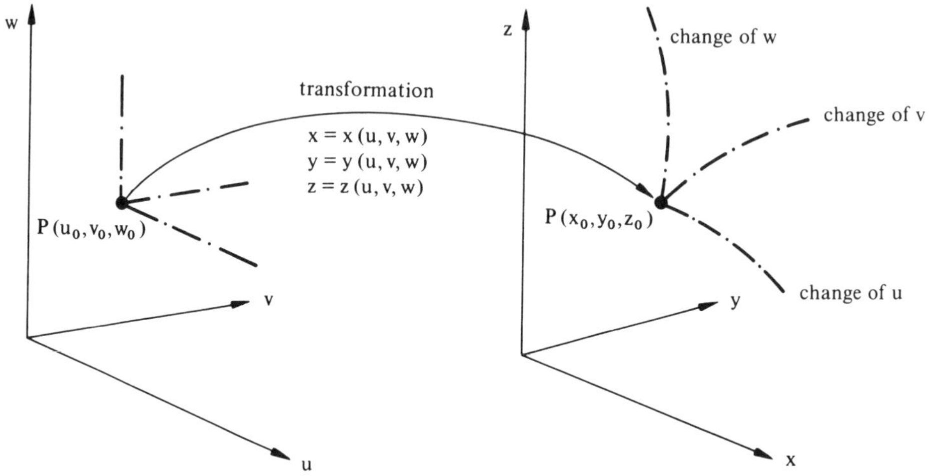

Fig. 12.1

is a tangent vector to the u curve, and for the unit vector $\mathbf{e}_u$ in the same direction, we have

$$\mathbf{e}_u = \frac{\dfrac{\partial \mathbf{r}}{\partial u}}{\left|\dfrac{\partial \mathbf{r}}{\partial u}\right|}$$

The vector given by (12.3) can be written in the form:

$$\frac{\partial \mathbf{r}}{\partial u} = h_u \mathbf{e}_u \tag{12.4}$$

where

$$h_u = \left|\frac{\partial \mathbf{r}}{\partial u}\right| \tag{12.5}$$

Likewise, if $\mathbf{e}_v$ and $\mathbf{e}_w$ designate the unit vectors tangent to the v and w coordinate curves respectively, then

$$\frac{\partial \mathbf{r}}{\partial v} = h_v \, \mathbf{e}_v, \qquad \frac{\partial \mathbf{r}}{\partial w} = h_w \, \mathbf{e}_w \tag{12.6}$$

with

$$h_v = \left| \frac{\partial \mathbf{r}}{\partial v} \right|, \qquad h_w = \left| \frac{\partial \mathbf{r}}{\partial w} \right| \tag{12.7}$$

The quantitics h_u, h_v, and h_w are called ***scale factors***. The unit vectors $\mathbf{e}_u$, $\mathbf{e}_v$, and $\mathbf{e}_w$ are oriented in the direction of increasing parameters u, v, and w, respectively. Further, we adopt the convention that at any point P (x, y, z), the three unit vectors $\mathbf{e}_u$, $\mathbf{e}_v$ and $\mathbf{e}_w$ form a right-handed system. If this triad is orthogonal, i.e., if the vectors $\mathbf{e}_u$, $\mathbf{e}_v$ and $\mathbf{e}_w$ are mutually perpendicular:

$$\mathbf{e}_u \cdot \mathbf{e}_v = \mathbf{e}_u \cdot \mathbf{e}_w = \mathbf{e}_v \cdot \mathbf{e}_w = 0$$

we say that u, v, and w are ***orthogonal curvilinear coordinates***.

As an example, consider the spherical coordinates r, ϑ, φ (fig. 12.2). We have

$$x = r \sin\vartheta \cos\varphi$$

$$y = r \sin\vartheta \sin\varphi$$

$$z = r \cos\vartheta$$

with $r \geq 0$, $0 \leq \varphi \leq 2\pi$ and $0 \leq \vartheta \leq \pi$.

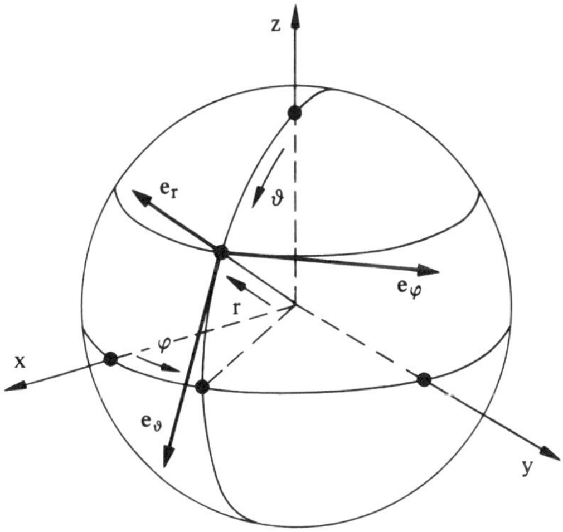

Fig. 12.2

For the scale factors, we immediately calculate

$$
\begin{aligned}
h_r &= 1 \\
h_\vartheta &= r \\
h_\varphi &= r \sin \vartheta
\end{aligned}
\tag{12.8}
$$

and verify that the unit vectors $\mathbf{e}_r$, $\mathbf{e}_\vartheta$, $\mathbf{e}_\varphi$ are orthogonal by pairs, i.e., that the parameters r, ϑ, and φ are orthogonal curvilinear coordinates.

12.2 DIFFERENTIAL ELEMENTS OF ARC LENGTH AND VOLUME

Let s be the arc length of a space curve described by the position vector $\mathbf{r} = \mathbf{r}(u, v, w)$. Then, the ***differential of the arc length*** ds is determined from the relation $ds^2 = d\mathbf{r} \cdot d\mathbf{r}$.

Because

$$d\mathbf{r} = \frac{\partial \mathbf{r}}{\partial u} du + \frac{\partial \mathbf{r}}{\partial v} dv + \frac{\partial \mathbf{r}}{\partial w} dw = h_u \mathbf{e}_u du + h_v \mathbf{e}_v dv + h_w \mathbf{e}_w dw$$

we have for ***orthogonal curvilinear coordinates***:

$$ds^2 = d\mathbf{r} \cdot d\mathbf{r} = h_u^2 du^2 + h_v^2 dv^2 + h_w^2 dw^2 \tag{12.9}$$

since

$$\mathbf{e}_u \cdot \mathbf{e}_v = \mathbf{e}_u \cdot \mathbf{e}_w = \mathbf{e}_v \cdot \mathbf{e}_w = 0$$

Taking into account the results (12.8) for a sphere, we find for the square of the differential ds^2 in ***spherical coordinates***:

$$ds^2 = dr^2 + r^2 \sin^2 \vartheta \, d\varphi^2 + r^2 \, d\vartheta^2$$

The ***differential element of volume*** dV is given by the volume of the parallelepiped determined by the three vectors (fig. 12.3):

$$\frac{\partial \mathbf{r}}{\partial u} du, \quad \frac{\partial \mathbf{r}}{\partial v} dv, \quad \frac{\partial \mathbf{r}}{\partial w} dw$$

from which we obtain for the volume dV:

$$dV = \left| \left(\frac{\partial \mathbf{r}}{\partial u} \wedge \frac{\partial \mathbf{r}}{\partial v} \right) \cdot \frac{\partial \mathbf{r}}{\partial w} \right| du \, dv \, dw \tag{12.10}$$

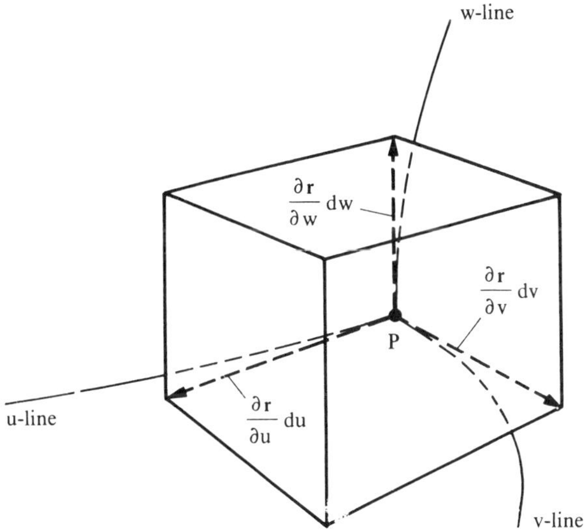

Fig. 12.3

We can easily see that this triple mixed product is equal to the determinant of a 3×3 matrix with rows $\partial \mathbf{r}/\partial u, \partial \mathbf{r}/\partial v$ and $\partial \mathbf{r}/\partial w$, respectively, such that

$$dV = \left\| \begin{matrix} \frac{\partial x}{\partial u} & \frac{\partial y}{\partial u} & \frac{\partial z}{\partial u} \\ \frac{\partial x}{\partial v} & \frac{\partial y}{\partial v} & \frac{\partial z}{\partial v} \\ \frac{\partial x}{\partial w} & \frac{\partial y}{\partial w} & \frac{\partial z}{\partial w} \end{matrix} \right\| du\, dv\, dw \tag{12.11}$$

This determinant is called the ***Jacobian*** of x, y, z, with respect to u, v, w, and is written

$$\frac{\partial(x, y, z)}{\partial(u, v, w)}$$

We, therefore, have

$$dV = \frac{\partial(x, y, z)}{\partial(u, v, w)} du\, dv\, dw \tag{12.12}$$

The differential element of volume in ***orthogonal curvilinear coordinates*** is a right prism formed by the three orthonormal vectors $\mathbf{e}_u$, $\mathbf{e}_v$, and $\mathbf{e}_w$. Consequently, formula (12.10) simplifies to

$$dV = h_u\, h_v\, h_w\, du\, dv\, dw \tag{12.13}$$

since

$$|(\mathbf{e}_u \wedge \mathbf{e}_v) \cdot \mathbf{e}_w| = 1$$

Finally, let us note that the differential element of volume in ***spherical coordinates*** is given by

$$dV = r^2 \sin\vartheta\, dr\, d\vartheta\, d\varphi$$

12.3 GRADIENT, DIVERGENCE, AND CURL IN ORTHOGONAL CURVILINEAR COORDINATES

Let $\phi(u, v, w)$ be a scalar function of the orthogonal curvilinear coordinates u, v, w, and let $\mathbf{F}(u, v, w) = F_u(u, v, w)\,\mathbf{e}_u + F_v(u, v, w)\,\mathbf{e}_v + F_w(u, v, w)\,\mathbf{e}_w$ be a vector field expressed in the moving triad $\mathbf{e}_u, \mathbf{e}_v, \mathbf{e}_w$. Then, always on the condition that the functions under consideration are sufficiently differentiable, we have the following results:

$$\textbf{grad}\,\phi = \nabla\phi = \frac{1}{h_u}\frac{\partial\phi}{\partial u}\mathbf{e}_u + \frac{1}{h_v}\frac{\partial\phi}{\partial v}\mathbf{e}_v + \frac{1}{h_w}\frac{\partial\phi}{\partial w}\mathbf{e}_w \tag{12.14}$$

$$\text{div}\,\mathbf{F} = \nabla\cdot\mathbf{F}$$

$$= \frac{1}{h_u h_v h_w}\left[\frac{\partial}{\partial u}(h_v h_w F_u) + \frac{\partial}{\partial v}(h_u h_w F_v) + \frac{\partial}{\partial w}(h_u h_v F_w)\right] \tag{12.15}$$

$$\textbf{curl}\,\mathbf{F} = \nabla\wedge\mathbf{F} = \frac{1}{h_u h_v h_w}\begin{vmatrix} h_u\,\mathbf{e}_u & h_v\,\mathbf{e}_v & h_w\,\mathbf{e}_w \\ \dfrac{\partial}{\partial u} & \dfrac{\partial}{\partial v} & \dfrac{\partial}{\partial w} \\ h_u F_u & h_v F_v & h_w F_w \end{vmatrix} \tag{12.16}$$

$$\text{div}\,\textbf{grad}\,\phi = \Delta\phi = \frac{1}{h_u h_v h_w}\left[\frac{\partial}{\partial u}\left(\frac{h_v h_w}{h_u}\frac{\partial\phi}{\partial u}\right)\right.$$

$$\left. + \frac{\partial}{\partial v}\left(\frac{h_u h_w}{h_v}\frac{\partial\phi}{\partial v}\right) + \frac{\partial}{\partial w}\left(\frac{h_u h_v}{h_w}\frac{\partial\phi}{\partial w}\right)\right] \tag{12.17}$$

12.3.1 Demonstration of the gradient formula

In order to determine the gradient of the scalar function ϕ (u, v, w) in orthogonal curvilinear coordinates, we have recourse to the formula:

$$d\phi = \mathbf{grad}\ \phi \cdot d\mathbf{r} \tag{12.18}$$

with

$$d\mathbf{r} = \frac{\partial \mathbf{r}}{\partial u} du + \frac{\partial \mathbf{r}}{\partial v} dv + \frac{\partial \mathbf{r}}{\partial w} dw$$

$$= h_u \mathbf{e}_u du + h_v \mathbf{e}_v dv + h_w \mathbf{e}_w dw \tag{12.19}$$

Let us assume that **grad** ϕ can be expressed in curvilinear coordinates:

$$\mathbf{grad}\ \phi = F_u \mathbf{e}_u + F_v \mathbf{e}_v + F_w \mathbf{e}_w \tag{12.20}$$

where F_u, F_v, and F_w are the components of **grad** ϕ in the moving triad. On one hand, it follows from equations (12.19) and (12.20) from in orthogonal curvilinear coordinates:

$$d\phi = \mathbf{grad}\ \phi \cdot d\mathbf{r} = h_u F_u du + h_v F_v dv + h_w F_w dw \tag{12.21}$$

and on the other hand, we have for the total differential $d\phi$:

$$d\phi = \frac{\partial \phi}{\partial u} du + \frac{\partial \phi}{\partial v} dv + \frac{\partial \phi}{\partial w} dw \tag{12.22}$$

Comparing the two relations (12.21) and (12.22), we obtain

$$\frac{\partial \phi}{\partial u} = h_u F_u, \quad \frac{\partial \phi}{\partial v} = h_v F_v, \quad \frac{\partial \phi}{\partial \phi} = h_w F_w$$

with the result that the gradient is expressed in orthogonal curvilinear coordinates as

$$\mathbf{grad}\ \phi = \frac{1}{h_u} \frac{\partial \phi}{\partial u} \mathbf{e}_u + \frac{1}{h_v} \frac{\partial \phi}{\partial v} \mathbf{e}_v + \frac{1}{h_w} \frac{\partial \phi}{\partial w} \mathbf{e}_w$$

which constitutes formula (12.14).

12.3.2 Auxiliary formulas

Let us apply the del operator ∇ to the scalar function $\phi = u$. According to formula (12.14), we have

$$\mathbf{grad}\ u = \frac{\mathbf{e}_u}{h_u} \tag{12.23}$$

In an analogous fashion, we obtain

$$\mathbf{grad}\ v = \frac{\mathbf{e}_v}{h_v} \quad \text{and} \quad \mathbf{grad}\ w = \frac{\mathbf{e}_w}{h_w} \tag{12.24}$$

Because **curl grad** $\phi = \mathbf{0}$, we find the following identity:

$$\mathbf{curl}\left(\frac{\mathbf{e}_u}{h_u}\right) = \mathbf{curl}\left(\frac{\mathbf{e}_v}{h_v}\right) = \mathbf{curl}\left(\frac{\mathbf{e}_w}{h_w}\right) = \mathbf{0} \tag{12.25}$$

In ***orthogonal curvilinear coordinates***, the unit vectors $\mathbf{e}_u$, $\mathbf{e}_v$, and $\mathbf{e}_w$ are mutually perpendicular, so that:

$$\mathbf{e}_w = \mathbf{e}_u \wedge \mathbf{e}_v, \quad \mathbf{e}_u = \mathbf{e}_v \wedge \mathbf{e}_w, \quad \mathbf{e}_v = \mathbf{e}_w \wedge \mathbf{e}_u \tag{12.26}$$

Substituting the relations (12.23) and (12.24) in equations (12.26), we obtain

$$\begin{aligned} \mathbf{e}_w &= h_u\, h_v\, \mathbf{grad}\ u \wedge \mathbf{grad}\ v \\ \mathbf{e}_u &= h_v\, h_w\, \mathbf{grad}\ v \wedge \mathbf{grad}\ w \\ \mathbf{e}_v &= h_u\, h_w\, \mathbf{grad}\ w \wedge \mathbf{grad}\ u \end{aligned} \tag{12.27}$$

Let ϕ and ψ be two arbitrary scalar functions. From formula (7.34), we have

$$\operatorname{div}(\mathbf{grad}\ \phi \wedge \mathbf{grad}\ \psi) = 0 \tag{12.28}$$

and using relations (12.27), we find

$$\operatorname{div}\left(\frac{\mathbf{e}_w}{h_u\, h_v}\right) = \operatorname{div}\left(\frac{\mathbf{e}_u}{h_v\, h_w}\right) = \operatorname{div}\left(\frac{\mathbf{e}_v}{h_u\, h_w}\right) = 0 \tag{12.29}$$

12.3.3 Demonstration of the divergence formula

To calculate the divergence of the vector field $\mathbf{F}$ directly in orthogonal curvilinear coordinates, we assume again that $\mathbf{F}$ is given in curvilinear coordinates:

$$\mathbf{F} = F_u\, \mathbf{e}_u + F_v\, \mathbf{e}_v + F_w\, \mathbf{e}_w$$

Then,

$$\operatorname{div} \mathbf{F} = \operatorname{div}(F_u\, \mathbf{e}_u) + \operatorname{div}(F_v\, \mathbf{e}_v) + \operatorname{div}(F_w\, \mathbf{e}_w) \tag{12.30}$$

We now consider each term on the right side of equation (12.30) separately. For the first term, we write

$$\text{div}\,(F_u\,\mathbf{e}_u) = \text{div}\left(h_v\,h_w\,F_u\,\frac{\mathbf{e}_u}{h_v\,h_w}\right)$$

and applying formula (7.33), we obtain

$$\text{div}\,(F_u\,\mathbf{e}_u) = h_v\,h_w\,F_u\,\text{div}\left(\frac{\mathbf{e}_u}{h_v\,h_w}\right) + \frac{\mathbf{e}_u}{h_v\,h_w}\cdot\mathbf{grad}\,(h_v\,h_w\,F_u) \tag{12.31}$$

From formula (12.29), the first term on the right side of relation (12.31) vanishes identically, and using the formula for the gradient (12.14), we get

$$\text{div}\,(F_u\,\mathbf{e}_u) = \frac{1}{h_u\,h_v\,h_w}\,\frac{\partial}{\partial u}\,(h_v\,h_w\,F_u)$$

By treating the two other terms of equation (12.30) in the same way, we get the divergence formula (12.15).

Let us also note that formula (12.17) follows immediately from formulas (12.14) and (12.15).

12.3.4 Demonstration of the formula for the curl

As in the preceding section, we assume that **F** is given in curvilinear coordinates:

$$\mathbf{F} = F_u\,\mathbf{e}_u + F_v\,\mathbf{e}_v + F_w\,\mathbf{e}_w$$

Then,

$$\mathbf{curl}\,\mathbf{F} = \mathbf{curl}\,(F_u\,\mathbf{e}_u) + \mathbf{curl}\,(F_v\,\mathbf{e}_v) + \mathbf{curl}\,(F_w\,\mathbf{e}_w) \tag{12.32}$$

Applying formula (7.35) to the first term on the right side of (12.32), we obtain

$$\mathbf{curl}\,(F_u\,\mathbf{e}_u) = \mathbf{curl}\left(h_u\,F_u\,\frac{\mathbf{e}_u}{h_u}\right)$$

$$= \mathbf{grad}\,(h_u\,F_u) \wedge \frac{\mathbf{e}_u}{h_u} + h_u\,F_u\,\mathbf{curl}\left(\frac{\mathbf{e}_u}{h_u}\right) \tag{12.33}$$

From formula (12.25), the second term on the right side of relations (12.33) vanishes identically. There remains

$$\mathbf{curl}\,(F_u\,\mathbf{e}_u) = \mathbf{grad}\,(h_u\,F_u) \wedge \frac{\mathbf{e}_u}{h_u} \tag{12.34}$$

and, according to the gradient formula (12.14), we get

$$\mathbf{grad}\,(h_u F_u) = \frac{1}{h_u}\,\frac{\partial}{\partial u}(h_u F_u)\,\mathbf{e}_u + \frac{1}{h_v}\,\frac{\partial}{\partial v}(h_u F_u)\,\mathbf{e}_v + \frac{1}{h_w}\,\frac{\partial}{\partial w}(h_u F_u)\,\mathbf{e}_w \tag{12.35}$$

so that after substitution of (12.35) into (12.34), we get

$$\mathbf{curl}\,(F_u\,\mathbf{e}_u) = \frac{1}{h_u h_v h_w}\left[h_v \frac{\partial}{\partial w}(h_u F_u)\,\mathbf{e}_v - h_w \frac{\partial}{\partial v}(h_u F_u)\,\mathbf{e}_w\right]$$

We proceed analogously for the other two terms on the right side of equation (12.32), and finally by presenting the result in the form of a determinant, we find formula (12.16).

12.3.5 Example: Cylindrical coordinates

In cylindrical coordinates, the position vector **r** of a point P is given by

$$\mathbf{r} = \rho \cos\varphi\,\mathbf{i} + \rho \sin\varphi\,\mathbf{j} + z\mathbf{k}$$

For the vectors tangent to the ρ, φ and z coordinate curves, we find

$$\frac{\partial \mathbf{r}}{\partial \rho} = \cos\varphi\,\mathbf{i} + \sin\varphi\,\mathbf{j}$$

$$\frac{\partial \mathbf{r}}{\partial \varphi} = -\rho \sin\varphi\,\mathbf{i} + \rho \cos\varphi\,\mathbf{j}$$

$$\frac{\partial \mathbf{r}}{\partial z} = \mathbf{k}$$

such that we obtain for the scale factors:

$$h_\rho = \left|\frac{\partial \mathbf{r}}{\partial \rho}\right| = 1$$

$$h_\varphi = \left|\frac{\partial \mathbf{r}}{\partial \varphi}\right| = \rho$$

$$h_z = \left|\frac{\partial \mathbf{r}}{\partial z}\right| = 1$$

The unit vectors tangent to the coordinate curves are according to (12.14) and (12.6):

$$\mathbf{e}_\rho = \cos\varphi\,\mathbf{i} + \sin\varphi\,\mathbf{j}$$

$$\mathbf{e}_\varphi = -\sin\varphi\,\mathbf{i} + \cos\varphi\,\mathbf{j}$$

$$\mathbf{e}_z = \mathbf{k}$$

It is easily verified that $\mathbf{e}_\rho$, $\mathbf{e}_\varphi$, and $\mathbf{e}_z$ are perpendicular by pairs and, consequently, the cylindrical coordinates are orthogonal curvilinear coordinates.

The quantities **grad** ϕ, div **F**, **curl F** and $\Delta\phi$ are expressed in cylindrical coordinates according to formulas (12.14) to (12.17) in the following way:

$$\mathbf{grad}\,\phi = \frac{\partial\phi}{\partial\rho}\,\mathbf{e}_\rho + \frac{1}{\rho}\,\frac{\partial\phi}{\partial\varphi}\,\mathbf{e}_\varphi + \frac{\partial\phi}{\partial z}\,\mathbf{e}_z \tag{12.36}$$

$$\text{div}\,\mathbf{F} = \frac{1}{\rho}\left[\frac{\partial}{\partial\rho}(\rho\,F_\rho) + \frac{\partial F_\varphi}{\partial\varphi} + \frac{\partial}{\partial z}(\rho\,F_z)\right] \tag{12.37}$$

$$\mathbf{curl}\,\mathbf{F} = \frac{1}{\rho}\begin{vmatrix} \mathbf{e}_\rho & \rho\,\mathbf{e}_\varphi & \mathbf{e}_z \\ \dfrac{\partial}{\partial\rho} & \dfrac{\partial}{\partial\varphi} & \dfrac{\partial}{\partial z} \\ F_\rho & \rho\,F_\varphi & F_z \end{vmatrix} \tag{12.38}$$

$$\Delta\phi = \text{div}\,\mathbf{grad}\,\phi = \frac{1}{\rho}\,\frac{\partial}{\partial\rho}\left(\rho\,\frac{\partial\phi}{\partial\rho}\right) + \frac{1}{\rho^2}\,\frac{\partial^2\phi}{\partial\varphi^2} + \frac{\partial^2\phi}{\partial z^2} \tag{12.39}$$

To conclude this chapter, we wish to solve ***Laplace's equation*** $\Delta\phi = 0$ in cylindrical coordinates, with the hypothesis that ϕ depends only on the radius $\rho = \sqrt{x^2 + y^2}$: $\phi = \phi(\rho)$. From the last formula (12.39), we have, in this case,

$$\Delta\phi = \frac{1}{\rho}\,\frac{\partial}{\partial\rho}\left(\rho\,\frac{\partial\phi}{\partial\rho}\right) = 0$$

from which we obtain, upon integrating the first time:

$$\rho\,\frac{\partial\phi}{\partial\rho} = c_1$$

and upon integrating the second time, after separation of the variables:

$$\phi = c_1 \ln \rho + c_2$$

where c_1 and c_2 are integration constants.

Part III
Analytical Methods for Solving Differential Equations

Chapter 13

Fourier Series and Applications

13.1 PRELIMINARY CONSIDERATIONS

A real-valued function f(t) defined on the real axis is called ***periodic***, if there is a positive constant T, such that for any t:

$$f(t+T) = f(t) \tag{13.1}$$

The smallest value of $T > 0$ for which the condition (13.1) is satisfied is called the ***period*** of f(t).

It can be easily proved that the trigonometric functions:

$$\cos \frac{2\pi k}{T} t, \quad \sin \frac{2\pi k}{T} t, \quad k = 1, 2, 3, \dots \tag{13.2}$$

are periodic functions whose period divides T. If the variable t designates the time, then the factor ω_k:

$$\omega_k = \frac{2\pi k}{T}$$

is the ***angular frequency*** expressed in radians per second.

A periodic function f(t) of period T is completely determined if it is known in an interval [a, a + T] of length T, where a is any real number. The function f* (t), which is equal to f(t) in the interval [a, a + T], and which vanishes beyond this interval, is called a ***generating function*** of f(t) (fig. 13.1). Starting with f* (t) we can reconstruct f(t) by shifting f* (t) to the left and to the right by multiples of the period T, which is expressed by the formula:

$$f(t) = \sum_{k=-\infty}^{\infty} f^*(t + kT) \tag{13.3}$$

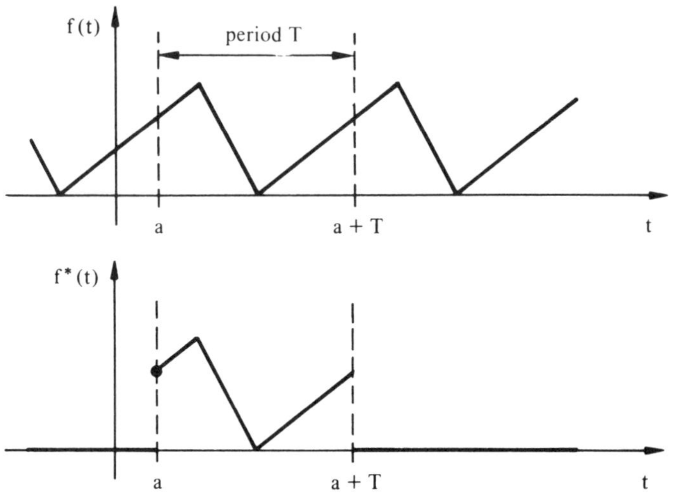

Fig. 13.1

A function f(t) is said to be ***piecewise continuous*** in a given interval, if this interval can be divided into a finite number of subintervals such that f(t) is continuous in each of them. The left-hand limit of f(t) at a point of discontinuity t_0 is written $f(t_0-0)$ and the right-hand limit at the same point $f(t_0+0)$. This situation is illustrated in figure 13.2.

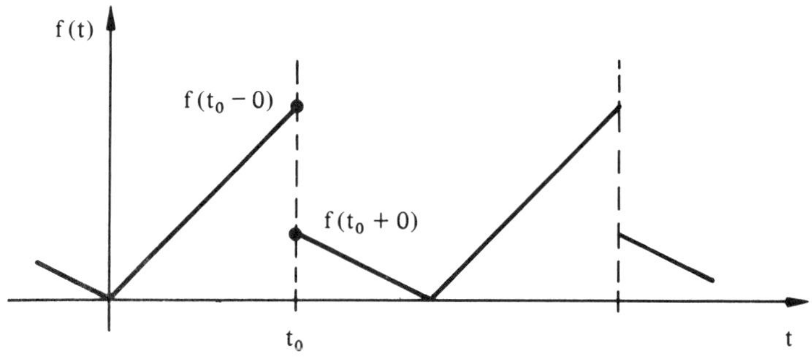

Fig. 13.2

Let us now consider the integral of a periodic function f(t) of period T over an arbitrary interval of length T, [a, a + T]:

$$\int_a^{a+T} f(t)\,dt = \int_a^T f(t)\,dt + \int_T^{a+T} f(t)\,dt \tag{13.4}$$

By substituting $t = \tau + T$ into the second integral on the right side of equation (13.4), we have

$$\int_a^{a+T} f(t)\,dt = \int_a^T f(t)\,dt + \int_0^a f(\tau + T)\,d\tau$$

and since $f(\tau + T) = f(\tau)$, we find:

$$\int_a^{a+T} f(t)\,dt = \int_0^T f(t)\,dt \tag{13.5}$$

i.e., the ***integration interval*** $[a, a+T]$ ***of a periodic function of period T*** can be replaced by any interval of length T, in particular $[0, T]$ or $[-T/2, T/2]$.

Hereafter, we shall use the formulas:

$$\int_0^T \cos\frac{2\pi k}{T}t \cdot \cos\frac{2\pi n}{T}t\,dt = \int_0^T \sin\frac{2\pi k}{T}t \cdot \sin\frac{2\pi n}{T}t\,dt = \begin{cases} 0 & \text{if } k \neq n \\ T/2 & \text{if } k = n \end{cases} \tag{13.6}$$

and

$$\int_0^T \cos\frac{2\pi k}{T}t \cdot \sin\frac{2\pi n}{T}t\,dt = 0 \tag{13.7}$$

where k and n can take on all integer values 1,2,3,. . .

We verify relations (13.6) and (13.7) from the following elementary formulas:

$$\cos\alpha\cos\beta = \frac{1}{2}\cos(\alpha-\beta) + \frac{1}{2}\cos(\alpha+\beta) \tag{13.8}$$

$$\sin\alpha\sin\beta = \frac{1}{2}\cos(\alpha-\beta) - \frac{1}{2}\cos(\alpha+\beta) \tag{13.9}$$

$$\sin\alpha\cos\beta = \frac{1}{2}\sin(\alpha-\beta) + \frac{1}{2}\sin(\alpha+\beta) \tag{13.10}$$

Relation (13.6) is established, for $k \neq n$, using

$$\int_0^T \cos \frac{2\pi (k \pm n)}{T} t \, dt = \int_0^T \sin \frac{2\pi (k \pm n)}{T} t \, dt = 0 \tag{13.11}$$

For $k = n$, i.e., for the integral of the square of the cosine or sine, we take into account $\cos^2\alpha = (1 + \cos 2\alpha)/2$ and $\sin^2\alpha = (1 - \cos 2\alpha)/2$ to obtain, using (13.11):

$$\int_0^T \cos^2 \frac{2\pi k}{T} t \, dt = \int_0^T \sin^2 \frac{2\pi k}{T} t \, dt = \frac{T}{2} \tag{13.12}$$

Finally, relation (13.7) is a consequence of the identity (13.11), which remains valid for the sine in the case where $k = n$.

13.2 FOURIER SERIES

The trigonometric functions (13.2):

$$\cos \frac{2\pi k}{T} t, \quad \sin \frac{2\pi k}{T} t, \quad k = 1, 2, 3, \ldots$$

are the simplest periodic functions. More complicated periodic functions are:

$$P_N (t) = \frac{a_0}{2} + \sum_{k=1}^{N} \left(a_k \cos \frac{2\pi k}{T} t + b_k \sin \frac{2\pi k}{T} t \right) \tag{13.13}$$

i.e., the linear combinations of the trigonometric functions (13.2), in which the constant term $a_0/2$ is modified to simplify the subsequent formulas. A function (13.13) is called ***a trigonometric polynomial of order*** N. This leads naturally to the problem of the approximation of an arbitrary periodic function $f(t)$ of period T by a trigonometric polynomial, and finally to that of the expansion of the function $f(t)$ in a ***trigonometric series***:

$$f(t) = \frac{a_0}{2} + \sum_{k=1}^{\infty} \left(a_k \cos \frac{2\pi k}{T} t + b_k \sin \frac{2\pi k}{T} t \right) \tag{13.14}$$

Thus, the problem consists of determining the coefficients:

$$a_0, a_1, b_1, a_2, b_2, \ldots$$

such that the series

$$\frac{a_0}{2} + \sum_{k=1}^{\infty} \left(a_k \cos \frac{2\pi k}{T} t + b_k \sin \frac{2\pi k}{T} t \right) \tag{13.15}$$

converges and that its sum is equal to the given periodic function f(t).

Let us assume that the series (13.15) converges to the periodic function f(t) of period T:

$$f(t) = \frac{a_0}{2} + \sum_{k=1}^{\infty} \left(a_k \cos \frac{2\pi k}{T} t + b_k \sin \frac{2\pi k}{T} t \right) \tag{13.16}$$

We wish to determine the coefficients $a_0, a_1, b_1, \ldots$ Integrating equation (13.16) term by term over the interval [0, T], we have

$$\int_0^T f(t)\, dt = \int_0^T \frac{a_0}{2}\, dt + \sum_{k=1}^{\infty} \left(a_k \int_0^T \cos \frac{2\pi k}{T} t\, dt + b_k \int_0^T \sin \frac{2\pi k}{T} t\, dt \right)$$

and, from (13.11), this implies the equality:

$$\int_0^T f(t)\, dt = \frac{a_0}{2} T$$

from which we obtain for the coefficient a_0:

$$a_0 = \frac{2}{T} \int_0^T f(t)\, dt \tag{13.17}$$

Next, we shall try to determine the coefficients a_k and b_k. Let n be an arbitrary positive integer. By multiplying the left- and right-hand sides of (13.16) by $\cos(2\pi nt/T)$, and integrating as before, we obtain

$$\int_0^T f(t) \cos \frac{2\pi n}{T} t\, dt = \frac{a_0}{2} \int_0^T \cos \frac{2\pi n}{T}\, dt$$

$$+ \sum_{k=1}^{\infty} \left(a_k \int_0^T \cos \frac{2\pi k}{T} t \cdot \cos \frac{2\pi n}{T} t\, dt + b_k \int_0^T \sin \frac{2\pi k}{T} t \cdot \cos \frac{2\pi n}{T} t\, dt \right) \tag{13.18}$$

From (13.6), (13.7), and (13.11), all the integrals of the right-hand side of relation (13.18) are zero, except one, which is the integral:

$$\int_0^T \cos \frac{2\pi k}{T} t \cdot \cos \frac{2\pi n}{T} t\, dt = \frac{T}{2} \quad \text{for } k = n$$

Relation (13.18) thus takes the form:

$$\int_0^T f(t) \cos \frac{2\pi n}{T} t\, dt = \frac{T}{2} a_n$$

from which we get

$$a_n = \frac{2}{T} \int_0^T f(t) \cos \frac{2\pi n}{T} t \, dt, \quad n = 0, 1, 2, \ldots \tag{13.19}$$

In the same way, we find the formula:

$$b_n = \frac{2}{T} \int_0^T f(t) \sin \frac{2\pi n}{T} t \, dt, \quad n = 1, 2, \ldots \tag{13.19'}$$

Summarizing, and writing k instead of n, we have

$$a_0 = \frac{2}{T} \int_0^T f(t) \, dt$$

$$a_k = \frac{2}{T} \int_0^T f(t) \cos \frac{2\pi k}{T} t \, dt, \quad k = 1, 2, \ldots \tag{13.20}$$

$$b_k = \frac{2}{T} \int_0^T f(t) \sin \frac{2\pi k}{T} t \, dt, \quad k = 1, 2, \ldots$$

The coefficients a_k and b_k are called ***Fourier coefficients*** of the given function f(t), and the series:

$$f(t) = \frac{a_0}{2} + \sum_{k=1}^{\infty} \left(a_k \cos \frac{2\pi k}{T} t + b_k \sin \frac{2\pi k}{T} t \right) \tag{13.21}$$

is called the ***Fourier series*** corresponding to f(t).

We have carried out the above considerations with the hypothesis that the given function allows for representation of (13.21), or, in other words, that it can be expanded in a convergent Fourier series. Further, it has been assumed that this series can be integrated term by term. Now, we wish to broach the problem in the following way: let there be a periodic function f(t). We determine its Fourier coefficients, and then examine if the corresponding Fourier series converges to f(t).

Before moving on to the solution of this problem, we wish to determine, as an example, the Fourier series representation of the periodic function $f(t) = \sin t$, $0 \leq t \leq \pi$, and which is repeated periodically outside the interval $[0, \pi]$. The function thus defined is an ***even*** function, which implies that its Fourier series does not contain ***any sine terms***. Since the period $T = \pi$, we have:

$$f(t) = \frac{a_0}{2} + \sum_{k=1}^{\infty} a_k \cos \frac{2\pi k}{T} t = \frac{a_0}{2} + \sum_{k=1}^{\infty} a_k \cos 2kt$$

and we calculate for the Fourier coefficients from (13.20):

$$a_0 = \frac{2}{\pi} \int_0^T \sin t \, dt = \frac{2}{\pi} (-\cos t)\Big|_0^{\pi} = \frac{4}{\pi}$$

$$a_k = \frac{2}{\pi} \int_0^T \sin t \cos 2kt \, dt$$

$$= \frac{1}{\pi} \int_0^T \sin (t + 2kt) \, dt + \frac{1}{\pi} \int_0^T \sin (t - 2kt) \, dt$$

$$= -\frac{1}{\pi} \left. \frac{\cos (t + 2kt)}{1 + 2k} \right|_0^{\pi} - \frac{1}{\pi} \left. \frac{\cos (t - 2kt)}{1 - 2k} \right|_0^{\pi}$$

$$= \frac{2}{\pi (1 + 2k)} + \frac{2}{\pi (1 - 2k)} = \frac{4}{\pi (1 - 4k^2)}$$

from which we obtain the Fourier series representation of the given function:

$$f(t) = \frac{2}{\pi} \left(1 + \sum_{k=1}^{\infty} \frac{2}{1 - 4k^2} \cos 2kt \right)$$

13.3 DIRICHLET'S THEOREM

The Fourier series of a periodic function $f(t)$ will be convergent and its sum will be equal to $f(t)$ if certain conditions are imposed on $f(t)$. ***Dirichlet's theorem***, a basic theorem in the theory of Fourier series, states: Let the function $f(t)$ be given in the interval of $[0, T]$ and defined beyond this interval by the relation $f(t) = f(t + T)$. If $f(t)$ and $f'(t)$ are piecewise continuous in the interval $[0, T]$, then the Fourier series of $f(t)$ converges to

- $f(t)$ if t is a point of continuity of $f(t)$;
- $(f(t+0) + f(t-0))/2$ if t is a point of discontinuity

The Dirichlet conditions imposed on the function $f(t)$ are sufficient, but not necessary, conditions. If these conditions are not satisfied, the Fourier series of $f(t)$ is not necessarily convergent. At present, there are no known necessary and sufficient conditions for convergence of Fourier series.

In order to simplify the demonstration of Dirichlet's theorem, we carry out the substitution $x = 2\pi t/T$ in the Fourier series of $f(t)$. The interval $[0, T]$ will thus be transformed to the interval $[0, 2\pi]$. We have

$$f(x) = \frac{a_0}{2} + \sum_{k=1}^{\infty} (a_k \cos kx + b_k \sin kx) \qquad (13.22)$$

where the Fourier coefficients are given by:

$$a_0 = \frac{1}{\pi} \int_0^{2\pi} f(x)\, dx$$

$$a_k = \frac{1}{\pi} \int_0^{2\pi} f(x) \cos kx\, dx, \quad k = 1, 2, \ldots \tag{13.23}$$

$$b_k = \frac{1}{\pi} \int_0^{2\pi} f(x) \sin kx\, dx, \quad k = 1, 2, \ldots$$

Let us now consider the trigonometric polynomial $P_N(x)$ of order N at a point $x = x_0$:

$$P_N(x_0) = \frac{a_0}{2} + \sum_{k=1}^{N} (a_k \cos k\, x_0 + b_k \sin k\, x_0) \tag{13.24}$$

By substituting the expressions for the Fourier coefficients into the right side of (13.24), we get, due to the trigonometric identity $\cos(\alpha - \beta) = \cos\alpha \cos\beta + \sin\alpha \sin\beta$:

$$P_N(x_0) = \frac{1}{2\pi} \int_0^{2\pi} f(x)\, dx + \frac{1}{\pi} \sum_{k=1}^{N}$$

$$\times \left(\int_0^{2\pi} f(x) \cos kx \cdot \cos kx_0\, dx + \int_0^{2\pi} f(x) \sin kx \cdot \sin kx_0\, dx \right)$$

$$= \frac{1}{2\pi} \int_0^{2\pi} f(x)\, dx + \frac{1}{\pi} \sum_{k=1}^{N} \int_0^{2\pi} f(x) \cos k(x - x_0)\, dx \tag{13.25}$$

Changing the sum of the integrals to the integral of the sum, we find for the polynomial (13.25):

$$P_N(x_0) = \frac{1}{\pi} \int_0^{2\pi} f(x)$$

$$\times \left\{ \frac{1}{2} + \cos(x - x_0) + \cos 2(x - x_0) + \ldots + \cos N(x - x_0) \right\} dx \tag{13.26}$$

In order to calculate the sum on the right side of (13.26), we consider the trigonometric relation:

$$\sin \frac{\varphi}{2} \cos k\varphi = \frac{1}{2} \sin\left(k + \frac{1}{2}\right)\varphi - \frac{1}{2} \sin\left(k - \frac{1}{2}\right)\varphi$$

For the sum of k = 1 to k = N of the differences on the right-hand side above, we obtain

$$\sin\frac{\varphi}{2}\,[\cos\varphi + \cos 2\varphi + \ldots + \cos N\varphi]$$

$$= \frac{1}{2}\left[\sin\frac{3\varphi}{2} - \sin\frac{\varphi}{2}\right] + \frac{1}{2}\left[\sin\frac{5\varphi}{2} - \sin\frac{3\varphi}{2}\right]$$

$$+ \ldots + \frac{1}{2}\left[\sin\left(N+\frac{1}{2}\right)\varphi - \sin\left(N-\frac{1}{2}\right)\varphi\right]$$

$$= \frac{1}{2}\sin\left(N+\frac{1}{2}\right)\varphi - \frac{1}{2}\sin\frac{\varphi}{2} \tag{13.27}$$

and dividing relation (13.27) by $\sin(\varphi/2)$ and adding 1/2 to both sides, we obtain the formula:

$$\frac{1}{2} + \cos\varphi + \cos 2\varphi + \ldots + \cos N\varphi = \frac{\sin\left(N+\frac{1}{2}\right)\varphi}{2\sin\frac{\varphi}{2}} \tag{13.28}$$

Thus, equation (13.26) is written in the following way:

$$P_N(x_0) = \frac{1}{2\pi}\int_0^{2\pi} f(x)\,\frac{\sin\left[\left(N+\frac{1}{2}\right)\left(x - x_0\right)\right]}{\sin\frac{x - x_0}{2}}\,dx \tag{13.29}$$

From formula (13.28), it is easily verified that

$$\int_0^{2\pi} \frac{\sin\left[\left(N+\frac{1}{2}\right)\left(x - x_0\right)\right]}{2\sin\frac{x - x_0}{2}}\,dx = \pi \tag{13.30}$$

and multiplying (13.30) on both sides by $f(x_0)$, we have

$$\frac{1}{2\pi}\int_0^{2\pi} f(x_0)\,\frac{\sin\left[\left(N+\frac{1}{2}\right)\left(x-x_0\right)\right]}{\sin\frac{x-x_0}{2}}\,dx = f(x_0) \tag{13.31}$$

Let us first examine the case in which the function f(x) and its derivative are continuous at any point x_0 of the real axis. To show that the Fourier series of f(x) converges at x_0 to $f(x_0)$, it must be shown that the difference $P_N(x_0) - f(x_0)$ approaches zero as N increases to infinity. With this aim, we subtract the identity (13.31) from the trigonometric polynomial (13.29):

$$P_N(x_0) - f(x_0) = \frac{1}{2\pi}\int_0^{2\pi} [f(x) - f(x_0)]\,\frac{\sin\left[\left(N+\frac{1}{2}\right)\left(x-x_0\right)\right]}{\sin\frac{x-x_0}{2}}\,dx \tag{13.32}$$

then consider the function $\psi(x)$:

$$\psi(x) = \frac{f(x) - f(x_0)}{\sin\frac{x-x_0}{2}} = \frac{f(x) - f(x_0)}{x - x_0}\cdot\frac{x-x_0}{\sin\frac{x-x_0}{2}}$$

in the interval $[0,2\pi]$. This function is clearly continuous for any $x \neq x_0$, and its behavior at the point $x = x_0$ is determined by the relation:

$$\lim_{x\to x_0} \frac{x-x_0}{\sin\frac{x-x_0}{2}} = 2$$

i.e., $\psi(x)$ converges to $2f'(x_0)$ as x approaches x_0. Thus, the function $\psi(x)$ is continuous in the entire interval $[0,2\pi]$. We must still evaluate the difference (13.32):

$$P_N(x_0) - f(x_0) = \frac{1}{2\pi}\int_0^{2\pi} \psi(x)\sin\left[\left(N+\frac{1}{2}\right)\left(x-x_0\right)\right]dx \tag{13.33}$$

Integrating by parts, it is easily seen that the integral on the right side of (13.33) approaches zero as $N\to\infty$ (i.e., the integral of a rapidly oscillating function approaches zero as its frequency is increased); this result is sometimes called Riemann's theorem. Thus, $P_N(x_0) - f(x_0)$ approaches zero for $N\to\infty$, and Dirichlet's theorem has been demonstrated for the case where f(x) and f′(x) are continuous.

The most general result in the case where f (x) and f′ (x) are piecewise continuous is easy to obtain. In this case, we substract the identity:

$$\frac{f(x_0+0)+f(x_0-0)}{2}=\frac{1}{2\pi}\int_0^{2\pi}\frac{f(x_0+0)+f(x_0-0)}{2}\cdot\frac{\sin\left[\left(N+\frac{1}{2}\right)\left(x-x_0\right)\right]}{\sin\frac{x-x_0}{2}}\,dx$$

from the relation (13.29) which gives

$$P_N(x_0)-\frac{f(x_0+0)+f(x_0-0)}{2}$$

$$=\frac{1}{4\pi}\int_0^{2\pi}[f(x)-f(x_0+0)]\,\frac{\sin\left[\left(N+\frac{1}{2}\right)\left(x-x_0\right)\right]}{\sin\frac{x-x_0}{2}}\,dx$$

$$+\frac{1}{4\pi}\int_0^{2\pi}[f(x)-f(x_0+0)]\,\frac{\sin\left[\left(N+\frac{1}{2}\right)\left(x-x_0\right)\right]}{\sin\frac{x-x_0}{2}}\,dx \qquad (13.34)$$

Because the integration interval for a periodic function of period 2π can be replaced by any interval of length 2π, we shall integrate the first term on the right side of (13.34) over the interval $[x_0, x_0+2\pi]$ and the second term over the interval $[x_0-2\pi, x_0]$. The function:

$$\psi(x)=\frac{f(x)-f(x_0+0)}{\sin\frac{x-x_0}{2}}$$

is piecewise continuous in $[x_0, x_0+2\pi]$; it converges to $2f'(x_0+0)$ as x approaches x_0 from the right. Likewise, the function:

$$\varphi(x)=\frac{f(x)-f(x_0-0)}{\sin\frac{x-x_0}{2}}$$

is piecewise continuous in $[x_0 - 2\pi, x_0]$; it converges toward $2f'(x_0 - 0)$ from the left. As Riemann's theorem, mentioned above, remains valid for piecewise continuous functions, the left side of (13.34) approaches zero, as $N \to \infty$. Dirichlet's theorem is thus demonstrated.

If the periodic function f(t) satisfies the conditions of Dirichlet's theorem, we have the following relation, called ***Parseval's identity***:

$$\frac{2}{T}\int_0^T f^2(t)\,dt = \frac{a_0^2}{2} + \sum_{k=1}^{\infty}(a_k^2 + b_k^2) \tag{13.35}$$

where a_k and b_k are the Fourier coefficients of f(t).

Without going into the details of the demonstration, we arrive at this identity (13.35) by multiplying both sides of relation (13.21) by $2f(t)/T$, and then integrating term by term over the interval [0, T]:

$$\frac{2}{T}\int_0^T f^2(t)\,dt = \frac{a_0}{2}\cdot\frac{2}{T}\int_0^T f(t)\,dt$$

$$+ \sum_{k=1}^{\infty}\left[a_k\cdot\frac{2}{T}\int_0^T f(t)\cos\frac{2\pi k}{T}t\,dt + b_k\cdot\frac{2}{T}\int_0^T f(t)\sin\frac{2\pi k}{T}t\,dt\right]$$

and by using relations (13.20) for the Fourier coefficients we obtain Parseval's identity (13.35).

13.4 COMPLEX NOTATION FOR FOURIER SERIES

Using Euler's identities:

$$e^{\pm j\varphi} = \cos\varphi \pm j\sin\varphi$$

the Fourier series (13.21) for f(t) can be written as

$$f(t) = \frac{a_0}{2} + \sum_{k=1}^{\infty}\frac{a_k - j\,b_k}{2}e^{j\frac{2\pi k}{T}t} + \sum_{k=1}^{\infty}\frac{a_k + j\,b_k}{2}e^{-j\frac{2\pi k}{T}t}$$

$$= \sum_{k=-\infty}^{\infty} c_k\, e^{j\frac{2\pi k}{T}} \tag{13.36}$$

where we have defined the ***complex Fourier coefficients*** as

$$c_k = \begin{cases} \dfrac{a_k - j\,b_k}{2}, & k = 1, 2, \ldots \\[2ex] \dfrac{a_0}{2}, & k = 0 \\[2ex] \dfrac{a_{-k} + j\,b_{-k}}{2}, & k = -1, -2, \ldots \end{cases}$$

For the complex Fourier coefficients c_k we obtain, according to formulas (13.20):

$$c_k = \begin{cases} \frac{1}{T}\int_0^T f(t)\left\{\cos\frac{2\pi k}{T}t - j\sin\frac{2\pi k}{T}t\right\}dt = \frac{1}{T}\int_0^T f(t)\, e^{-j\frac{2\pi k}{T}t}\,dt, & k = 1, 2, \ldots \\ \frac{1}{T}\int_0^T f(t)\,dt, \quad k = 0 & \\ \frac{1}{T}\int_0^T f(t)\left\{\cos\frac{-2\pi k}{T}t + j\sin\frac{-2\pi k}{T}t\right\}dt = \frac{1}{T}\int_0^T f(t)\, e^{-j\frac{2\pi k}{T}t}\,dt, & k = -1, -2, \ldots \end{cases}$$

which can be reduced to the single formula:

$$c_k = \frac{1}{T}\int_0^T f(t)\, e^{-j\omega_k t}\,dt, \quad k = 0, \pm 1, \pm 2, \ldots$$

In ***complex notation***, the ***Fourier series*** of a periodic function f(t) of period T, thus, takes the following form:

$$f(t) = \sum_{k=-\infty}^{\infty} c_k\, e^{j\omega_k t} \tag{13.37}$$

with the ***complex Fourier coefficients*** given by:

$$c_k = \frac{1}{T}\int_0^T f(t)\, e^{-j\omega_k t}\,dt \tag{13.38}$$

where

$$\omega_k = \frac{2\pi k}{T} \tag{13.39}$$

The complex coefficients c_k constitute the ***spectrum*** of the periodic function f(t). If we graph $|c_k|$ as a function of ω_k, we obtain the ***amplitude spectrum*** of f(t), and if we graph the argument of c_k as a function of ω_k, we have the ***phase spectrum*** of f(t). Because all the angular frequencies ω_k are multiples of the lowest frequency $\omega_1 = 2\pi/T$, we say that ω_1 is the ***fundamental angular frequency*** and that ω_k $(k \neq 1)$ is the ***kth harmonic frequency***.

As an example, let us determine the spectrum of a periodic train of rectangular pulses illustrated in figure 13.3.

The expansion in a complex Fourier series of f(t) is given by (13.37):

$$f(t) = \sum_{k=-\infty}^{\infty} c_k\, e^{j\omega_k t}, \quad \omega_k = \frac{2\pi k}{T}$$

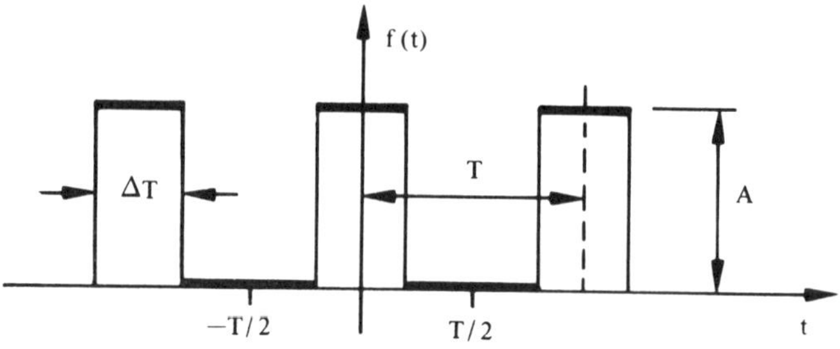

Fig. 13.3

The integration interval in the calculation of the Fourier coefficients (13.38) can be replaced by any other interval of the same length, namely by $[-T/2, T/2]$ to exploit the symmetry of the given function. We calculate

$$c_k = \frac{1}{T}\int_{-T/2}^{T/2} f(t).e^{-j\omega_k t}\,dt = \frac{1}{T}\int_{-\Delta T/2}^{\Delta T/2} A\,e^{-j\omega_k t}\,dt$$

$$= \frac{A}{T}\,\frac{e^{-j\omega_k \Delta T/2} - e^{j\omega_k \Delta T/2}}{-j\omega_k} = \frac{2A}{T\omega_k}\sin(\omega_k\,\Delta T/2)$$

which can be written in the following form:

$$c_k = \frac{A\,\Delta T}{T}\,\frac{\sin x}{x}, \qquad x = \omega_k\,\Delta T/2 \tag{13.40}$$

Because the coefficients c_k are real, we trace c_k directly as a function of ω_k (fig. 13.4). We obtain the zeros enclosing the principal lobe of the spectral envelope by setting $x = \pi$ in $\sin x/x$, from which we get $\omega\,\Delta T/2 = \pi$, which gives $\omega = 2\pi/\Delta T$.

Because $c_k = c_{-k}$, i.e., the complex Fourier coefficients are symmetrical, the expansion in a complex Fourier series of the periodic pulse train reduces to the real Fourier series representation of an even function:

$$f(t) = \frac{A\,\Delta T}{T}\sum_{k=-\infty}^{\infty}\frac{\sin(\omega_k\,\Delta T/2)}{\omega_k\,\Delta T/2}\,e^{j\omega_k t}$$

$$= \frac{A\,\Delta T}{T}\left\{1 + 2\sum_{k=1}^{\infty}\frac{\sin(\omega_k\,\Delta T/2)}{\omega_k\,\Delta T/2}\cos\omega_k t\right\}$$

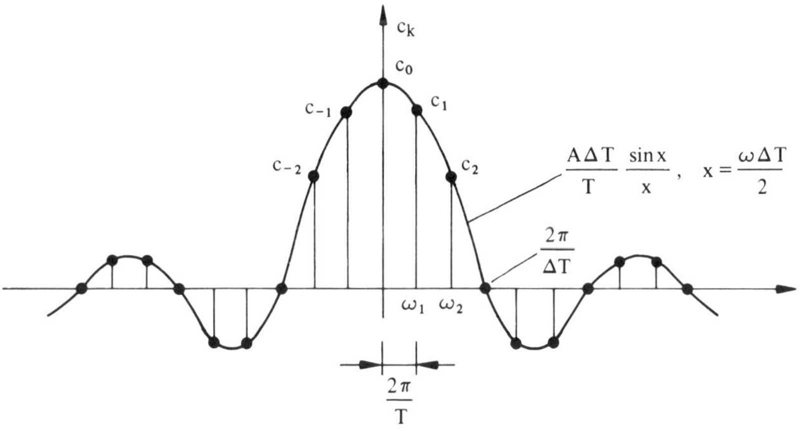

Fig. 13.4

13.5 TECHNICAL APPLICATIONS

13.5.1 Example: R-C circuit

In order to demonstrate the usefulness of Fourier series, we shall consider the electrical R-C circuit of figure 13.5. The input voltage v_e is related to the current i across the resistance R and capacitor C by the relation:

$$v_e(t) = Ri(t) + \frac{1}{C}\left\{ \int_0^t i(\tau)\, d\tau + Q_0 \right\} \tag{13.41}$$

where Q_0 is the initial charge of the capacitor C. The voltage at the output v_s is given by

$$v_s(t) = \frac{1}{C}\left\{ \int_0^t i(\tau)\, d\tau + Q_0 \right\} \tag{13.42}$$

from which we obtain for the current i, by differentiating equation (13.42):

$$i(t) = C\,\dot{v}_s(t) \tag{13.43}$$

Substituting the current (13.43) into equation (13.41), we obtain the diffferential equation:

$$v_e = RC\,\dot{v}_s + v_s \tag{13.44}$$

subject to the initial condition $v_s(0) = Q_0/C$.

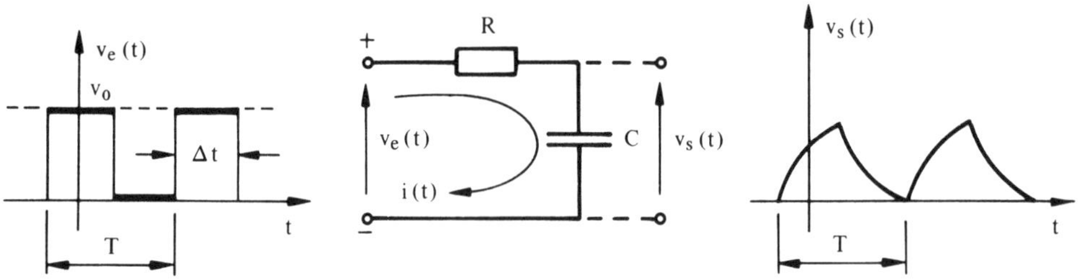

Fig. 13.5

Let us assume that the input voltage v_e is given by a periodic train of rectangular pulses (which we have just studied in section 13.4). Because v_e is a periodic function, we set, in complex notation (13.37) and (13.39):

$$v_e(t) = \sum_{k=-\infty}^{\infty} c_k\, e^{j\omega_k t}, \qquad \omega_k = \frac{2\pi k}{T} \tag{13.45}$$

The solution of the differential equation (13.44) is given by the sum of the solution of the associated homogeneous equation (i.e., by $\alpha \exp(-t/RC)$ and a particular solution of (13.44). The homogeneous solution is called the ***transient response*** of the circuit since it vanishes as t increases toward infinity. It is assumed that the particular solution is also a periodic function of the same period T as $v_e(t)$. Thus, we set

$$v_s(t) = \alpha\, e^{-\frac{t}{RC}} + \sum_{k=-\infty}^{\infty} c_k^*\, e^{j\omega_k t} \tag{13.46}$$

where the second term on the right side of equation (13.46) constitutes the ***steady-state response*** of the circuit for the given periodic excitation. Substituting v_s from (13.46) into the differential equation (13.44), we have

$$\sum_{k=-\infty}^{\infty} c_k\, e^{j\omega_k t} = R\,C \sum_{k=-\infty}^{\infty} j\omega_k\, c_k^*\, e^{j\omega_k t} + \sum_{k=-\infty}^{\infty} c_k^*\, e^{j\omega_k t} \tag{13.47}$$

Because the functions $\exp(j\,\omega_k t)$ are linearly independent, equation (13.47) must be satisfied for each integer k, from which we get

$$c_k = j\,R\,C\,\omega_k\, c_k^* + c_k^*$$

We, therefore, find

$$c_k^* = \frac{c_k}{1 + j\omega_k\, R\,C}$$

with (see formula (13.40)):

$$c_k = \frac{v_0 \Delta T}{T} \frac{\sin(\omega_k \Delta T/2)}{\omega_k \Delta T/2}$$

i.e., it is found that each component of the spectrum of the input signal c_k is multiplied by $1/(1 + j\omega_k RC)$. We set

$$H(j\omega) = \frac{1}{1 + j\omega RC} \tag{13.48}$$

and call this function $H(j\omega)$ the ***frequency response*** of the circuit; its absolute value:

$$|H(j\omega)| = \frac{1}{\sqrt{1 + \omega^2 R^2 C^2}} \tag{13.49}$$

is called the ***amplitude response*** of the circuit, while its argument:

$$\arg H(j\omega) = -\arctan(\omega R C) \tag{13.50}$$

is called the ***phase response*** of the circuit. The two functions are graphed in figure 13.6.

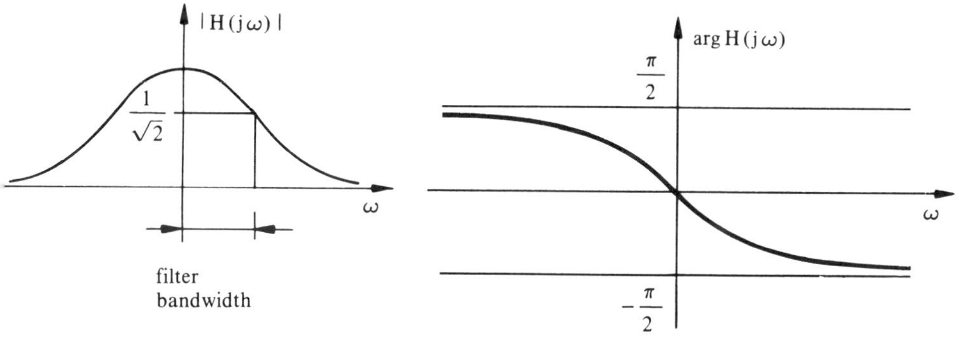

Fig. 13.6

13.5.2 Example: Heat equation

Consider a square plate with sides of length π. We assume that the faces of this plate are insulated, with three of its sides maintained at zero temperature and the fourth at a temperature T_0. The problem consists of determining the temperature of the ***equilibrium state*** at any point of the plate. If the xy coordinate system of figure 13.7 is chosen, the partial differential equation which governs the temperature $u(x,y,t)$ at the point (x,y) at the instant t is the following:

$$\frac{\partial u}{\partial t} = \kappa \left(\frac{\partial^2 u}{\partial x^2} + \frac{\partial^2 u}{\partial y^2} \right)$$

where $\kappa \neq 0$ is a constant. Because we are interested in the temperature of the equilibrium state, we set $\partial u / \partial t = 0$, which implies that the temperature satisfies ***Laplace's equation*** in two variables:

$$\frac{\partial^2 u}{\partial x^2} + \frac{\partial^2 u}{\partial y^2} = 0 \tag{13.51}$$

subject to the boundary conditions:

$$u(0, y) = u(x, 0) = u(\pi, y) = 0, \qquad u(x, \pi) = T_0$$

To solve equation (13.51), we set $u(x,y) = g(x)\,h(y)$. With this substitution, equation (13.51) is transformed into

$$g''h + gh'' = 0$$

from which we get

$$\frac{g''(x)}{g(x)} = -\frac{h''(y)}{h(y)} \tag{13.52}$$

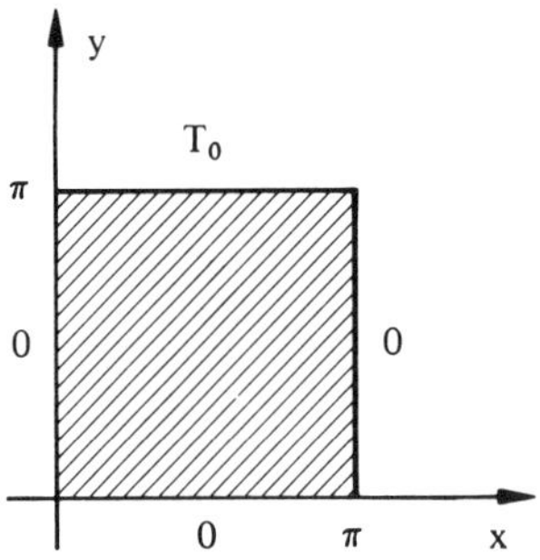

Fig. 13.7

Since the left side of equation (13.52) is a function of x, while the right side is a function of y, the two sides are necessarily equal to a constant, namely $-\lambda^2$. Thus, we obtain

$$g''(x) + \lambda^2 g(x) = 0$$

and

$$h''(y) - \lambda^2 h(y) = 0$$

By integrating these two second-order differential equations, we find

$$g(x) = a_1 \cos \lambda x + b_1 \sin \lambda x$$

$$h(y) = a_2 \cosh \lambda y + b_2 \sinh \lambda y$$

A possible solution of (13.51) is, thus,

$$u(x, y) = (a_1 \cos \lambda x + b_1 \sin \lambda x)(a_2 \cosh \lambda y + b_2 \sinh \lambda y)$$

The boundary conditions $u(0,y) = 0$ and $u(x,0) = 0$ imply, respectively, that $a_1 = 0$ and $a_2 = 0$, and the condition $u(\pi,y) = 0$ gives

$$u(\pi, y) = b_1 \sin \lambda \pi \cdot b_2 \sinh \lambda y = 0$$

To obtain a nontrivial solution, i.e., $b_1 \neq 0$ and $b_2 \neq 0$, we must choose $\lambda = k$, $k = 1,2,$ ..., from which we finally get

$$u_k(x, y) = B_k \sin kx \cdot \sinh ky, \quad k = 1, 2, \ldots$$

To satisfy the last condition $u(x,\pi) = T_0$, we first use the superposition theorem to obtain the general solution $u(x,y)$:

$$u(x, y) = \sum_{k=1}^{\infty} u_k(x, y) = \sum_{k=1}^{\infty} B_k \sin kx \cdot \sinh ky \tag{13.53}$$

Then, based on the boundary condition:

$$u(x, \pi) = \sum_{k=1}^{\infty} B_k \sinh \pi k \cdot \sin kx = T_0 \tag{13.54}$$

we determine the coefficients B_k in the following way. Equation (13.54) represents the Fourier series expansion of the constant T_0 in the interval $[0, \pi]$. Setting $u(x,\pi) = -T_0$ for $x \in [-\pi,0]$, the function $u(x,\pi)$ is defined in the entire interval $[-\pi,\pi]$, and since it is an odd function, we have for its Fourier series expansion:

$$u(x, \pi) = \sum_{k=1}^{\infty} b_k \sin kx \tag{13.55}$$

where

$$b_k = \frac{1}{\pi} \int_{-\pi}^{\pi} u(x, \pi) \sin kx \, dx, \quad k = 1, 2, \ldots \tag{13.56}$$

Because $u(x,\pi)$ and $\sin kx$ are odd functions, their product is an even function, and the integral (13.56) becomes

$$b_k = \frac{2}{\pi}\int_0^{\pi} u(x,\pi)\sin kx\,dx = \frac{2\,T_0}{\pi}\int_0^{\pi}\sin kx\,dx$$

$$= \frac{2T_0}{\pi}\left.\frac{-\cos kx}{k}\right|_0^{\pi} = \frac{2T_0}{\pi k}(1-\cos k\pi)$$

Comparing the two Fourier series (13.54) and (13.55), we find for the coefficients B_k:

$$B_k = \frac{2T_0\,(1-\cos k\pi)}{k\,\pi\,\sinh k\pi}$$

Finally, substituting these expressions for B_k in (13.53), we arrive at the solution:

$$u(x,y) = \frac{2T_0}{\pi}\sum_{k=1}^{\infty}\frac{(1-\cos k\pi)}{k\sinh k\pi}\sin kx\cdot\sinh ky \tag{13.57}$$

Chapter 14

Fourier Transform and Applications

14.1 NONPERIODIC FUNCTIONS

Since many practical problems do not involve periodic functions, it is desirable to generalize the Fourier series method to nonperiodic functions by letting the period approach infinity.

Let us first consider the periodic function f(t) of period T, satisfying the conditions of Dirichlet's theorem, with the following Fourier-series representation:

$$f(t) = \sum_{k=-\infty}^{\infty} c_k \, e^{j\omega_k t}, \quad \omega_k = 2\pi k/T \tag{14.1}$$

with

$$c_k = \frac{1}{T} \int_{-T/2}^{T/2} f(t) \, e^{-j\omega_k t} \, dt \tag{14.2}$$

Defining the function $F_T(\omega)$ by the integral:

$$F_T(\omega) = \int_{-T/2}^{T/2} f(t) \, e^{-j\omega t} \, dt \tag{14.3}$$

we have for the Fourier coefficients:

$$c_k = \frac{1}{T} F_T(\omega_k) \tag{14.4}$$

and the Fourier series (14.1) becomes

$$f(t) = \frac{1}{T} \sum_{k=-\infty}^{\infty} F_T(\omega_k) \, e^{j\omega_k t}, \quad \omega_k = 2\pi k/T \tag{14.5}$$

We note by $\Delta\omega$ the interval between two successive harmonics:

$$\Delta\omega = 2\pi/T \tag{14.6}$$

from which we obtain, substituting $\Delta\omega$ into the Fourier series (14.5):

$$f(t) = \frac{1}{2\pi} \sum_{k=-\infty}^{\infty} F_T(\omega_k)\, e^{j\omega_k t}\, \Delta\omega \tag{14.7}$$

If the period T approaches infinity, the function $F_T(\omega)$ approaches a limit $F(\omega)$:

$$F(\omega) = \lim_{T\to\infty} F_T(\omega) = \int_{-\infty}^{\infty} f(t)\, e^{-j\omega t}\, dt \tag{14.8}$$

The series (14.7) represents the sum of the areas of the rectangles of width $\Delta\omega$ and height $F_T(\omega_k)\exp(j\omega_k t)$. As $T\to\infty$, the interval $\Delta\omega$ between two harmonics approaches zero and, consequently, the limit of the sum (14.7) for $\Delta\omega \to 0$ equals the integral:

$$f(t) = \lim_{\Delta\omega\to 0} \frac{1}{2\pi} \sum_{k=-\infty}^{\infty} F_T(\omega_k)\, e^{j\omega_k t}\, \Delta\omega = \frac{1}{2\pi} \int_{-\infty}^{\infty} F(\omega)\, e^{j\omega t}\, d\omega \tag{14.9}$$

It must be noted here that the final stage of our consideration is not rigorous. In fact, for a rigorous demonstration of the validity of relations (14.8) and (14.9), it would be necessary to assume that f(t) is ***absolutely integrable*** over $(-\infty, \infty)$, i.e.,

$$\int_{-\infty}^{\infty} |f(t)|\, dt < \infty \tag{14.10}$$

Relation (14.9) shows that a nonperiodic function satisfying the condition (14.10) is obtained by synthesis of an infinitely large number of sinusoidal components $\exp(j\omega t)$ having infinitely small amplitudes $F(\omega)\,d\omega/2\pi$. This is why the function $F(\omega)$ is called the ***spectral density*** of f(t), and the integral (14.9) constitutes the ***spectral integral*** of f(t). The angular frequency ω of these components takes on all values of the real axis $(-\infty, \infty)$.

14.2 FOURIER TRANSFORM

According to the considerations discussed in the preceding section, we can establish the following ***theorem***.

Let f(t) be a real-valued function defined on the real axis, such that f(t) and f′(t) are piecewise continuous in any finite interval, and such that

$$\int_{-\infty}^{\infty} |f(t)|\, dt < \infty$$

i.e., it is assumed that f (t) is absolutely integrable on the real axis $(-\infty, \infty)$. Thus, at any point of continuity of f (t), we have

$$f(t) = \frac{1}{2\pi} \int_{-\infty}^{\infty} F(\omega)\, e^{j\omega t}\, d\omega \tag{14.11}$$

where

$$F(\omega) = \int_{-\infty}^{\infty} f(t)\, e^{-j\omega t}\, dt \tag{14.12}$$

At the points of discontinuity of f (t), we must replace f (t) in (14.11) by the average value $(f(t+0)+f(t-0))/2$ of the left- and right-hand limits of f (t) at that point.

The function F (ω) defined in (14.12) is called the ***Fourier transform*** of f (t) (or spectral density of f (t)), while f (t) in (14.11) is called the ***inverse Fourier transform*** of F (ω) (or spectral integral of f (t)). The pair of the two transforms (14.11) and (14.12) comprise the ***Fourier integral formulas.***

The fact that F (ω) is the Fourier transform of f (t) is written in the following manner:

$$f(t) \rightarrow F(\omega)$$

It is also noted that $F(\omega) = \mathscr{F}(f(t))$, respectively $f(t) = \mathscr{F}^{-1}(F(\omega))$.

As an example, let us determine the Fourier transform of the function:

$$f(t) = \begin{cases} 1 \text{ for } |t| \leq 1 \\ 0 \text{ for } |t| > 1 \end{cases}$$

According to the definition (13.12), we have

$$F(\omega) = \int_{-1}^{1} e^{-j\omega t}\, dt = \left.\frac{e^{-j\omega t}}{-j\omega}\right|_{-1}^{1}$$

$$= \frac{e^{-j\omega} - e^{j\omega}}{-j\omega} = 2\frac{\sin\omega}{\omega}$$

From the theorem on Fourier integrals mentioned above, the function f (t) can be represented at any point $|t| \neq 1$ by its inverse Fourier transform (14.11):

$$f(t) = \frac{1}{2\pi} \int_{-\infty}^{\infty} 2\frac{\sin\omega}{\omega} e^{j\omega t}\, d\omega$$

The points $|t| = 1$ are points of discontinuity of f (t), and at these points we have

$$\frac{1}{2} = \frac{1}{\pi} \int_{-\infty}^{\infty} \frac{\sin \omega}{\omega} e^{\pm j\omega} \, d\omega$$

Taking the sum of the two last integrals for t = 1 and t = –1, we obtain

$$1 = \frac{1}{\pi} \int_{-\infty}^{\infty} \frac{\sin \omega}{\omega} (e^{j\omega} + e^{-j\omega}) \, d\omega$$

$$= \frac{2}{\pi} \int_{-\infty}^{\infty} \frac{\sin \omega}{\omega} \cos \omega \, d\omega$$

$$= \frac{1}{\pi} \int_{-\infty}^{\infty} \frac{\sin 2\omega}{\omega} \, d\omega$$

and after changing the integration variable to $x = 2\omega$, we get

$$\int_{-\infty}^{\infty} \frac{\sin x}{x} \, dx = \pi \qquad (14.13)$$

The graphs of the two functions f(t) and F(ω) are given in figure 14.1.

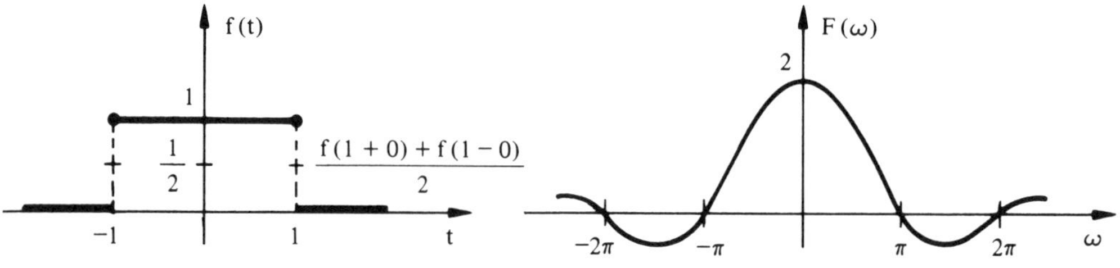

Fig. 14.1

14.3 PROPERTIES OF THE FOURIER TRANSFORM

It is assumed that the functions f(t) and g(t) considered in this section are real-valued functions, which are absolutely integrable over (– ∞, ∞).

In the following we shall consider some of the important results involving Fourier transforms and corresponding inverse Fourier transforms.

14.3.1 Linearity

It follows immediately from definition (14.12) of the Fourier transform that if

$$f(t) \rightarrow F(\omega)$$
$$g(t) \rightarrow G(\omega)$$

then

$$\lambda f(t) + \mu g(t) \rightarrow \lambda F(\omega) + \mu G(\omega) \tag{14.14}$$

where λ and μ are any constant factors.

14.3.2 Convolution

Let us assume further that f(t) and g(t) are ***square integrable*** over $(-\infty, \infty)$, i.e.,

$$\int_{-\infty}^{\infty} f^2(t)\,dt < \infty \quad \text{and} \quad \int_{-\infty}^{\infty} g^2(t)\,dt < \infty$$

The function h(t) defined on the real axis by the integral:

$$h(t) = \int_{-\infty}^{\infty} f(\tau)\,g(t-\tau)\,d\tau = \int_{-\infty}^{\infty} f(t-\tau)\,g(\tau)\,d\tau \tag{14.15}$$

is called the ***convolution*** of the two functions f(t) and g(t).

Thus, the Fourier transform $H(\omega)$ of h(t) equals

$$H(\omega) = \int_{-\infty}^{\infty} h(t)\,e^{-j\omega t}\,dt = \int_{-\infty}^{\infty} \left\{ \int_{-\infty}^{\infty} f(\tau)\,g(t-\tau)\,d\tau \right\} e^{-j\omega t}\,dt$$
$$= \int_{-\infty}^{\infty} \left\{ \int_{-\infty}^{\infty} f(\tau)\,g(t-\tau)\,e^{-j\omega(t-\tau)}\,e^{-j\omega\tau}\,d\tau \right\} dt$$

After changing the order of integration, we get

$$H(\omega) = \int_{-\infty}^{\infty} f(\tau)\,e^{-j\omega\tau} \left\{ \int_{-\infty}^{\infty} g(t-\tau)\,e^{-j\omega(t-\tau)}\,dt \right\} d\tau \tag{14.16}$$

Carrying out the substitution $t^* = t - \tau$ in the inner integral on the right side of relation (14.16), it follows that $H(\omega) = F(\omega)\,G(\omega)$, where $F(\omega)$ and $G(\omega)$ designate the Fourier transforms of f(t) and g(t), respectively. In summary,

$$h(t) = \int_{-\infty}^{\infty} f(\tau)\, g(t-\tau)\, d\tau \rightarrow H(\omega) = F(\omega)\, G(\omega) \qquad (14.17)$$

i.e., the Fourier transform of a convolution of two functions is equal to the product of their Fourier transforms.

14.3.3 Derivative

Let $F(\omega)$ be the Fourier transform of $f(t)$ and $G(\omega)$ the Fourier transform of the derivative $df(t)/dt$:

$$G(\omega) = \int_{-\infty}^{\infty} \frac{df(t)}{dt} e^{-j\omega t}\, dt$$

By hypothesis, $f(t)$ is absolutely integrable over the interval $(-\infty, \infty)$. This implies that $f(t)$ approaches zero as t approaches infinitely. Integrating by parts, we get

$$G(\omega) = f(t)\, e^{-j\omega t} \Big|_{-\infty}^{\infty} + j\omega \int_{-\infty}^{\infty} f(t)\, e^{-j\omega t}\, dt$$

$$= j\omega \int_{-\infty}^{\infty} f(t)\, e^{-j\omega t}\, dt$$

as the exponential function $\exp(-j\omega t)$ has the absolute value one, and $f(t) \rightarrow 0$ as $t \rightarrow \pm\infty$. The following relation results:

$$\frac{df(t)}{dt} \rightarrow j\omega F(\omega) \qquad (14.18)$$

14.3.4 First translation theorem

Let $F(\omega)$ be the Fourier transform of $f(t)$ and $G(\omega)$ is the Fourier transform of $\exp(j\omega_0 t) f(t)$, i.e.,

$$G(\omega) = \int_{-\infty}^{\infty} f(t)\, e^{j\omega_0 t}\, e^{-j\omega t}\, dt$$

$$= \int_{-\infty}^{\infty} f(t)\, e^{-j(\omega - \omega_0)t}\, dt$$

We, therefore, have

$$e^{j\omega_0 t} f(t) \rightarrow F(\omega - \omega_0) \qquad (14.19)$$

14.3.5 Second translation theorem

Let $g(t) = f(t - \tau)$, where τ designates a real constant. Thus, the Fourier transform $G(\omega)$ of $g(t)$ equals

$$G(\omega) = \int_{-\infty}^{\infty} g(t)\, e^{-j\omega t}\, dt = \int_{-\infty}^{\infty} f(t-\tau)\, e^{-j\omega t}\, dt$$

Setting $t - \tau = \lambda$, we get

$$G(\omega) = \int_{-\infty}^{\infty} f(\lambda)\, e^{-j\omega(\tau+\lambda)}\, d\lambda$$

$$= e^{-j\omega\tau} \int_{-\infty}^{\infty} f(\lambda)\, e^{-j\omega\lambda}\, d\lambda$$

from which we obtain the following rule:

$$f(t-\tau) \rightarrow e^{-j\omega\tau} F(\omega) \qquad (14.20)$$

14.4 APPLICATIONS OF THE FOURIER TRANSFORM

14.4.1 Example: Heat propagation in an infinite bar

We shall consider an infinite bar which coincides with the x-axis. In such a bar, the propagation of heat is governed by the partial differential equation:

$$\frac{\partial u}{\partial t} = \kappa \frac{\partial^2 u}{\partial x^2} \qquad (14.21)$$

where u designates the temperature, t is the time, and κ is a constant. The temperature $u(x,t)$ satisfies the partial differential equation (14.21) in the half-plane $-\infty < x < \infty$, $t \geq 0$ and also the boundary condition:

$$u(x, 0) = f(x) \qquad (14.22)$$

where $f(x)$ is the distribution of the temperature across the bar at the instant $t = 0$.

We shall find the solution $u(x,t)$ of this problem by using the Fourier transform method. To this end, we consider the Fourier transforms with respect to the variable x on the left and right sides of equation (14.21):

$$\int_{-\infty}^{\infty} e^{-j\omega x} \frac{\partial u}{\partial t}(x, t)\, dx = \kappa \int_{-\infty}^{\infty} e^{-j\omega x} \frac{\partial^2 u}{\partial x^2}(x, t)\, dx \qquad (14.23)$$

where t is treated as a parameter. Denoting the Fourier transform of u(x,t) by U(ω,t), we obtain from the differentiation rule (14.18) for (14.23):

$$\frac{\partial U}{\partial t}(\omega, t) = (j\omega)^2 \kappa U(\omega, t) = -\omega^2 \kappa U(\omega, t)$$

We integrate this last differential equation with respect to the variable t, from which we get

$$U(\omega, t) = C(\omega)\, e^{-\omega^2 \kappa t} \tag{14.24}$$

where the integration constant $C(\omega)$ is a function of ω. In order to determine $C(\omega)$, we use the boundary condition (14.22). For $t=0$, we find from (14.24) and the definition of the Fourier transform:

$$U(\omega, 0) = C(\omega) = \int_{-\infty}^{\infty} u(x, 0)\, e^{-j\omega x}\, dx$$

$$= \int_{-\infty}^{\infty} f(x)\, e^{-j\omega x}\, dx \tag{14.25}$$

As an example, let us determine the temperature in this bar for the initial temperature:

$$f(x) = \begin{cases} u_0 & \text{for } |x| \leq 1 \\ 0 & \text{for } |x| > 1 \end{cases}$$

We have, from (14.25):

$$C(\omega) = u_0 \int_{-1}^{1} e^{-j\omega x}\, dx = 2 u_0 \frac{\sin \omega}{\omega}$$

such that, by virtue of (14.24):

$$U(\omega, t) = 2 u_0 \frac{\sin \omega}{\omega} e^{-\omega^2 \kappa t}$$

Finally, the temperature u(x,t) is the inverse Fourier transform of U(ω,t), given by

$$u(x, t) = \frac{u_0}{\pi} \int_{-\infty}^{\infty} \frac{\sin \omega}{\omega} e^{-\omega^2 \kappa t} e^{j\omega x}\, d\omega$$

This situation is graphed in figure 14.2.

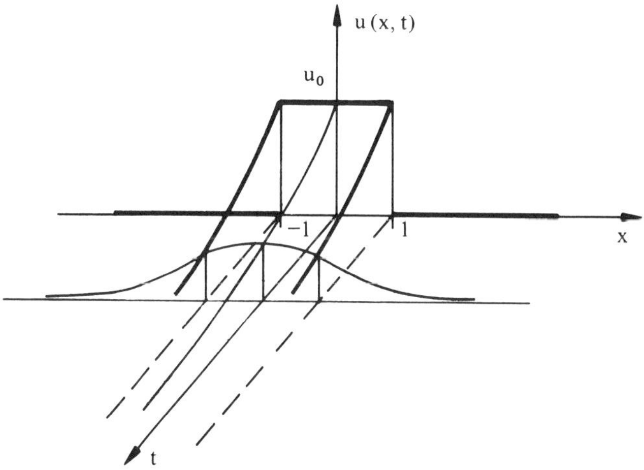

Fig. 14.2

14.4.2 Example: Dirac delta function

The Dirac delta function is a very important tool in technical applications. Its rigorous mathematical theory is based on that of generalized functions (distributions); in any case, a large number of its properties can be established directly.

To introduce the Dirac delta function, we start with the rectangular function $\Delta_\alpha(t)$ of duration $\alpha > 0$ and of amplitude $1/\alpha$, represented in figure 14.3.

Then, the ***Dirac delta function*** $\delta(t)$ is by definition the limit of the function $\Delta_\alpha(t)$, as the parameter α approaches zero, $\alpha > 0$:

$$\delta(t) = \lim_{\alpha \to 0} \Delta_\alpha(t) \tag{14.26}$$

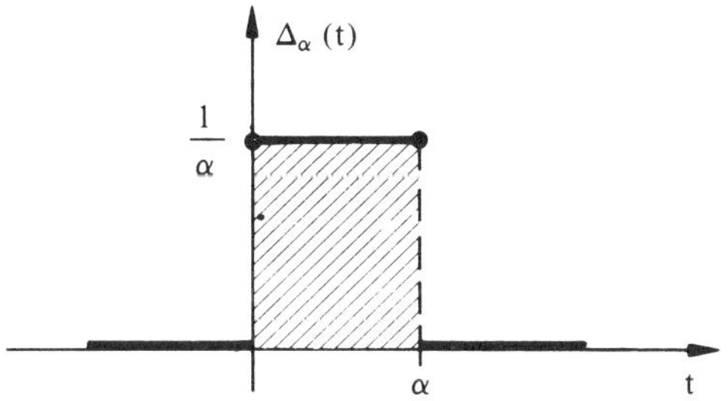

Fig. 14.3

From this definition, we immediately deduce the following properties of the function $\delta(t)$:

- it takes the values:

$$\delta(t) = \begin{cases} 0 & \text{for } t \neq 0 \\ \infty & \text{for } t = 0 \end{cases} \tag{14.27}$$

- its integral over the real axis is equal to one:

$$\int_{-\infty}^{\infty} \delta(t)\, dt = 1 \tag{14.28}$$

- for any function f(t) it possesses the sifting property:

$$\int_{-\infty}^{\infty} f(\tau)\, \delta(t-\tau)\, d\tau = f(t) \tag{14.29}$$

at any point of continuity of f(t).

We demonstrate this last property in the following way. According to the definition of $\delta(t)$, we have

$$\int_{-\infty}^{\infty} f(\tau)\, \delta(t-\tau)\, d\tau = \lim_{\alpha \to 0} \frac{1}{\alpha} \int_{t-\alpha}^{t} f(\tau)\, d\tau$$

and using the mean-value theorem of the integral calculus:

$$\int_{t-\alpha}^{t} f(\tau)\, d\tau = \alpha f(\xi), \qquad t - \alpha \leq \xi \leq t$$

we see that any point of continuity of f(t), as α approaches zero:

$$\int_{-\infty}^{\infty} f(\tau)\, \delta(t-\tau)\, d\tau = \lim_{\alpha \to 0} f(\xi) = f(t)$$

Using the sifting property (14.29), we can easily obtain the Fourier transform of $\delta(t)$. In effect, the integral (14.29) is a convolution. Let $F(\omega)$ and $G(\omega)$ be the Fourier transform of f(t) and $\delta(t)$ respectively. From property (14.17) of the Fourier transform, we have

$$F(\omega)\, G(\omega) = F(\omega)$$

from which we get

$$G(\omega) = 1$$

i.e., the ***Fourier transform*** of the ***Dirac delta function equals one***.

This last result can be verified directly from the definition of $\delta(t)$:

$$G(\omega) = \lim_{\alpha \to 0} \int_{\infty}^{\infty} \Delta_\alpha(t)\, e^{-j\omega t}\, dt$$

and an intermediate calculation gives

$$\int_{-\infty}^{\infty} \Delta_\alpha(t)\, e^{-j\omega t}\, dt = \int_0^{\alpha} \frac{1}{\alpha} e^{-j\omega t}\, dt = \frac{1}{-j\omega\alpha} e^{-j\omega t} \Big|_0^{\alpha} = \frac{1 - e^{-j\omega\alpha}}{j\omega\alpha}$$

$$= \frac{e^{-j\omega\alpha/2}}{\omega\alpha/2} \; \frac{e^{j\omega\alpha/2} - e^{-j\omega\alpha/2}}{2j} = e^{-j\omega\alpha/2} \frac{\sin(\omega\alpha/2)}{\omega\alpha/2}$$

The result is that

$$G(\omega) = \lim_{\alpha \to 0} e^{-j\omega\alpha/2} \frac{\sin(\omega\alpha/2)}{\omega\alpha/2} = 1$$

14.4.3 Example: Correlation function

Let us assume that the functions $f(t)$ and $g(t)$ are absolutely and square integrable over $(-\infty, \infty)$.

The correlation function $\varphi_{fg}(t)$ of the two functions $f(t)$ and $g(t)$ is defined by the integral:

$$\varphi_{fg}(t) = \int_{-\infty}^{\infty} f(\tau)\, g(t + \tau)\, d\tau \qquad (14.30)$$

If the spectral densities of $f(t)$ and $g(t)$ are denoted by $F(\omega)$ and $G(\omega)$, respectively, we have for the correlation function (14.30), using the inverse Fourier transform (14.11) of $g(t)$:

$$\varphi_{fg}(t) = \int_{-\infty}^{\infty} f(\tau) \left\{ \frac{1}{2\pi} \int_{-\infty}^{\infty} G(\omega)\, e^{j\omega(t+\tau)}\, d\omega \right\} d\tau$$

and after changing the order of integration:

$$\varphi_{fg}(t) = \frac{1}{2\pi} \int_{-\infty}^{\infty} \left\{ \int_{-\infty}^{\infty} f(\tau)\, e^{j\omega\tau}\, d\tau \right\} G(\omega)\, e^{j\omega t}\, d\omega$$

According to definition (14.12) of the Fourier transform of f(t), we get, taking account of the fact that f(t) and g(t) are real-valued functions:

$$\varphi_{fg}(t) = \frac{1}{2\pi} \int_{-\infty}^{\infty} \overline{F}(\omega)\, G(\omega)\, e^{j\omega t}\, d\omega$$

where $\overline{F}(\omega)$ designates the complex conjugate of $F(\omega)$. Setting $t = 0$, we find

$$\int_{-\infty}^{\infty} f(\tau)\, g(\tau)\, d\tau = \frac{1}{2\pi} \int_{-\infty}^{\infty} \overline{F}(\omega)\, G(\omega)\, d\omega \tag{14.31}$$

and if, in particular, $f(t) = g(t)$, we arrive at ***Parseval's identity***:

$$\int_{-\infty}^{\infty} f^2(t)\, dt = \frac{1}{2\pi} \int_{-\infty}^{\infty} |F(\omega)|^2\, d\omega \tag{14.32}$$

As an example, we shall calculate the integral:

$$\int_{-\infty}^{\infty} \frac{\sin^2 \omega}{\omega^2}\, d\omega$$

using identity (14.32).

As the Fourier transform $f(\omega)$ of the function:

$$f(t) = \begin{cases} 1 \text{ for } |x| \leq 1 \\ 0 \text{ for } |x| > 1 \end{cases}$$

equals

$$F(\omega) = \int_{-1}^{1} e^{-j\omega t}\, dt = \frac{2 \sin \omega}{\omega}$$

we obtain from (14.32):

$$\frac{1}{2\pi} \int_{-\infty}^{\infty} \frac{4 \sin^2 \omega}{\omega^2}\, d\omega = \int_{-1}^{1} dt = 2$$

Thus,

$$\int_{\infty}^{\infty} \frac{\sin^2 \omega}{\omega^2}\, d\omega = \pi$$

Chapter 15

Laplace Transform and Applications

15.1 DEFINITION

Let f(t) be a real-valued function, which is defined for all positive values of the real variable t ($t \geq 0$). Further, it is assumed that the function f(t) is piecewise continuous in the interval $0 \leq t < \infty$. To assure the existence of certain integrals, we impose the restriction on f(t), that there are real constants $M > 0$ and σ_0, such that

$$|f(t)| \leq M e^{\sigma_0 t} \tag{15.1}$$

for any value of t taken in the interval $0 \leq t < \infty$. The smallest value of σ_0 for which the inequality (15.1) is satisfied is called the ***abscissa of convergence*** of the Laplace transform of f(t).

Let us now consider the improper integral:

$$\int_0^{\infty} e^{-pt} f(t)\, dt \tag{15.2}$$

where $p = \sigma + j\omega$ is a complex variable. If the function f(t) satifies condition (15.1) and $\text{Re}\, p = \sigma > \sigma_0$, then the convergence of the integral (15.2) is ensured:

$$\left| \int_0^{\infty} e^{-pt} f(t) dt \right| = \left| \int_0^{\infty} e^{-\sigma t} e^{-j\omega t} f(t)\, dt \right|$$

$$\leq \int_0^{\infty} e^{-\sigma t} |f(t)|\, dt \leq M \int_0^{\infty} e^{-(\sigma - \sigma_0)t} dt = \frac{M}{\sigma - \sigma_0}, \quad \sigma > \sigma_0 \tag{15.3}$$

as $|\exp(j\omega t)| = 1$. Thus, the integral (15.2) exists, and defines a certain function of p, which is denoted by F(p):

$$F(p) = \int_0^{\infty} e^{-pt} f(t)\, dt \tag{15.4}$$

This function F(p) is called the single-sided ***Laplace transform*** of f(t). Furthermore the original function f(t) is called the ***inverse transform*** of F(p). The fact that F(p) is the transform of the function f(t) is expressed in the following way:

$$f(t) \rightarrow F(p)$$

The Laplace transform is also denoted by $F(p) = \mathscr{L}(f(t))$, and its inverse by $f(t) = \mathscr{L}^{-1}(F(p))$.

It follows immediately from the definition (15.4) that the Laplace transform is a ***linear*** operation. In effect, if the functions f(t) and g(t) have Laplace tranforms:

$$f(t) \rightarrow F(p)$$
$$g(t) \rightarrow G(p)$$

then,

$$\lambda f(t) + \mu g(t) \rightarrow \lambda F(p) + \mu G(p) \tag{15.5}$$

where λ and μ are any constant factors.

15.2 LAPLACE TRANSFORM OF SOME ELEMENTARY FUNCTIONS

15.2.1 Unit step function

The function u(t) thus defined:

$$u(t) = \begin{cases} 1 \text{ for } t \geq 0 \\ 0 \text{ for } t < 0 \end{cases} \tag{15.6}$$

is called the ***unit step*** function. The graph of the unit step function is shown in figure 15.1.

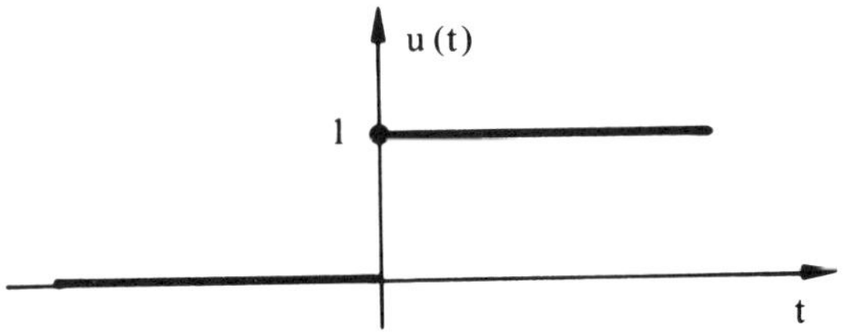

Fig. 15.1

Thus, according to definition (15.4), the Laplace transform U(p) of u(t) equals

$$U(p) = \int_0^\infty e^{-pt}\,dt = -\frac{e^{-pt}}{p}\Bigg|_0^\infty = \frac{1}{p} \tag{15.7}$$

provided that Re $p > 0$.

15.2.2 Exponential and trigonometric functions

Let $f(t) = e^{-\alpha t}$, $0 \le t < \infty$. Then,

$$F(p) = \int_0^\infty e^{-\alpha t} e^{-pt}\,dt = -\frac{e^{-(\alpha+p)t}}{(\alpha+p)}\Bigg|_0^\infty = \frac{1}{\alpha+p} \tag{15.8}$$

if Re $p > -\text{Re}\,\alpha$.

From this result, the Laplace transform of sin ωt follows immediately. In effect,

$$\sin \omega t = \frac{e^{j\omega t} - e^{-j\omega t}}{2j}$$

from which we obtain, by using the linearity of the Laplace transform (15.5):

$$\sin \omega t \rightarrow \frac{1}{2j}\left\{\frac{1}{p-j\omega} - \frac{1}{p+j\omega}\right\} = \frac{\omega}{p^2+\omega^2} \tag{15.9}$$

if Re $p > 0$.

Likewise, we find the Laplace transform of cos ωt. We have,

$$\cos \omega t = \frac{e^{j\omega t} + e^{-j\omega t}}{2}$$

and, consequently,

$$\cos \omega t \rightarrow \frac{1}{2}\left\{\frac{1}{p-j\omega} + \frac{1}{p+j\omega}\right\} = \frac{p}{p^2+\omega^2} \tag{15.10}$$

if Re $p > 0$.

15.2.3 Dirac delta function

Starting with the definition of the Dirac delta function $\delta(t)$ (14.4.2):

$$\delta(t) = \lim_{\alpha \to 0} \Delta_\alpha(t)$$

where

$$\Delta_\alpha(t) = \begin{cases} 1/\alpha \text{ for } 0 \leq t \leq \alpha, & \alpha > 0 \\ 0 \text{ for t otherwise} \end{cases}$$

we have

$$\delta(t) \rightarrow \lim_{\alpha \to 0} \int_0^\alpha \frac{1}{\alpha} e^{-pt} dt = \lim_{\alpha \to 0} \frac{1 - e^{-p\alpha}}{\alpha p}$$

and expanding exp $(-p\alpha)$ in a Taylor series:

$$e^{-p\alpha} = 1 - p\alpha + \frac{p^2\alpha^2}{2} - \frac{p^3\alpha^3}{6} + \ldots$$

we find for the Laplace transform of $\delta(t)$:

$$\delta(t) \rightarrow \lim_{\alpha \to 0} \left\{ 1 - \frac{p\alpha}{2} + \frac{p^2\alpha^2}{6} - \ldots \right\} = 1 \tag{15.11}$$

15.3 THEOREMS ON LAPLACE TRANSFORMS

15.3.1 Laplace transform of the derivative and primitive

Let us assume that the function f(t) and all its derivatives $f'(t), f''(t), \ldots, f^{(n)}(t)$ satisfy condition (15.1).

Let F(p) be the Laplace transform of f(t). For the transform of f′(t), we have according to definition (15.4):

$$f'(t) \rightarrow \int_0^\infty e^{-pt} f'(t)\, dt$$

Integrating by parts, we obtain

$$\int_0^\infty e^{-pt} f'(t)\, dt = e^{-pt} f(t) \Big|_0^\infty + p \int_0^\infty e^{-pt} f(t)\, dt \tag{15.12}$$

However, f(t) satisfies condition (15.1), and, consequently, if Re $p > \sigma_0$:

$$\lim_{t \to \infty} e^{-pt} f(t) = 0$$

Thus, we find for the ***Laplace transform of the derivative*** of a function f(t):

$$f'(t) \rightarrow pF(p) - f(0) \tag{15.13}$$

Let us now consider the transform of the second derivative $f''(t)$. Substituting the expression $pF(p)-f(0)$ into formula (15.13) in place of $F(p)$, and replacing $f(t)$ with $f'(t)$, we have

$$f''(t) \rightarrow p(pF(p)-f(0))-f'(0)$$

which gives

$$f''(t) \rightarrow p^2 F(p) - pf(0) - f'(0) \tag{15.14}$$

In the same way, we find for the transform of the *n*th derivative $f^{(n)}(t)$:

$$f^{(n)}(t) \rightarrow p^n F(p) - p^{n-1} f(0) - p^{n-2} f'(0) - \ldots - f^{(n-1)}(0) \tag{15.15}$$

Next, let us consider the Laplace transform of the definite integral:

$$g(t) = \int_0^t f(\tau)\, d\tau$$

According to definition of the definite integral, we have $g'(t) = f(t)$, and taking the Laplace transform on the right and left sides of the preceding relation, we obtain, from formula (15.13):

$$pG(p) - g(0) = F(p)$$

and since $g(0) = 0$:

$$G(p) = \frac{F(p)}{p}$$

Therefore, we obtain for the ***Laplace transform of the primitive*** of a function $f(t)$:

$$\int_0^t f(\tau)\, dt \rightarrow \frac{F(p)}{p} \tag{15.16}$$

As an example, we consider the unit ramp $f(t) = t$, $t \geq 0$. This function is the definite integral of the unit step function $u(t)$. Because the transform of the function $u(t)$ equals $1/p$ (formula(15.7)), we find

$$t \rightarrow \frac{1}{p^2} \tag{15.17}$$

and, generally,

$$\frac{t^n}{n!} \rightarrow \frac{1}{p^{n+1}} \tag{15.18}$$

15.3.2 Translation theorems

If F(p) is the Laplace transform of the function f(t), then the transform of the function exp($-\alpha$t) f(t) if given by F(p + α):

$$e^{-\alpha t} f(t) \to \int_0^\infty e^{-\alpha t} f(t) e^{-pt} dt = \int_0^\infty f(t) e^{-(\alpha+p)t} dt = F(p+\alpha) \quad (15.19)$$

under the condition that Re(p + α)$>\sigma_0$, where σ_0 is the abscissa of convergence of F(p). This fact is known as the ***first translation theorem***.

To establish the ***second translation theorem***, we consider the delayed function $u(t-\tau)f(t-\tau)$, $\tau>0$, where u is the unit step function. It is identically zero for $t<\tau$. Then, according to the definition of the Laplace transform:

$$u(t-\tau) f(t-\tau) \to \int_\tau^\infty f(t-\tau) e^{-pt} dt.$$

Carrying out the change of variable $t-\tau=\eta$ in the preceding integral, we obtain

$$\int_\tau^\infty f(t \epsilon \tau) e^{-pt} dt = \int_0^\infty f(\eta) e^{-(\eta+\tau)p} d\eta = e^{-\tau p} \int_0^\infty f(\eta) e^{-\eta p} d\eta$$

Thus, we obtain

$$u(t-\tau) f(t-\tau) \to e^{-\tau p} F(p) \quad (15.20)$$

15.3.3 Convolution theorem

In the solution of differential equations by the Laplace transform method, the convolution of two functions is often used. Let f(t) and g(t) be two functions that vanish identically for $t<0$. Then, their ***convolution*** h(t) is defined by the integral:

$$h(t) = \int_0^t f(\tau) g(t-\tau) d\tau = \int_0^t f(t-\tau) g(\tau) d\tau \quad (15.21)$$

Starting with definition (15.4), we find for the Laplace transform H(p) of h(t):

$$H(p) = \int_0^\infty h(t) e^{-pt} dt = \int_0^\infty \left\{ \int_0^t f(\tau) g(t-\tau) d\tau \right\} e^{-pt} dt \quad (15.22)$$

This integral is a double integral, extending over the region limited by the lines $\tau=0$ and $\tau=t$ (fig. 15.2)

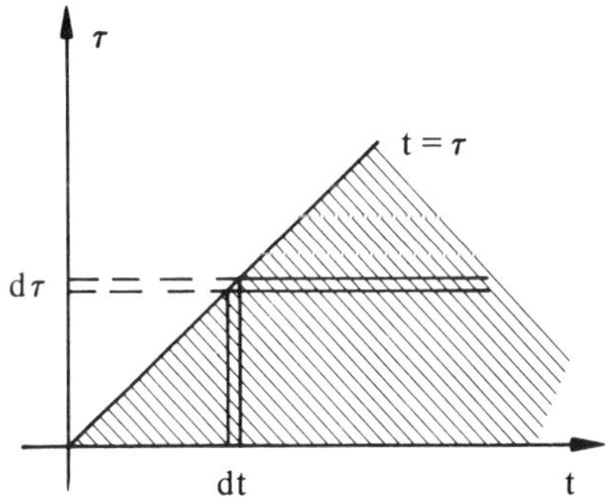

Fig. 15.2

Changing the order of integration in the integral (15.22), we obtain

$$H(p)=\int_0^\infty f(\tau)\left\{\int_\tau^\infty g(t-\tau)\,e^{-pt}\,dt\right\}d\tau$$

and carrying out the change of variable $t-\tau=\eta$ in the inner integral of this last double integral, we arrive at

$$H(p)=\int_0^\infty f(\tau)\left\{\int_0^\infty g(\eta)\;e^{-p(\eta+\tau)}\,d\eta\right\}d\tau$$

$$=\int_0^\infty f(\tau)\,e^{-p\tau}\,d\tau\int_0^\infty g(\eta)e^{-p\eta}\,d\eta=F(p)\,G(p)$$

Thus, we have found the ***convolution theorem***, which states:

$$\int_0^t f(\tau)\,g(t-\tau)\,d\tau\rightarrow F(p)\,G(p) \tag{15.23}$$

15.3.4 Final-value theorem

The ***final-value theorem*** states:

$$\lim_{t\to\infty} f(t)=\lim_{p\to 0}\,p\,F(p) \tag{15.24}$$

if the indicated limits exist.

To demonstrate this theorem, we use rule (15.13) for the transform of the derivative:

$$\int_0^\infty f'(t)\, e^{-pt}\, dt = p\,F(p) - f(0)$$

Letting p approach zero in the preceding equation, we obtain

$$\lim_{p \to 0} \int_0^\infty f'(t)\, e^{-pt}\, dt = \int_0^\infty f'(t)\, dt = \lim_{t \to \infty} f(t) - f(0)$$

from which we get relation (15.24).

15.3.5 Derivative of the transform

Let $f(t) \to F(p)$. Then, we have for the ***derivative of the transform***:

$$t\,f(t) \to -\frac{d\,F(p)}{dp} \tag{15.25}$$

Let us first note, without demonstration, that if f(t) satisfies condition (15.1), the transform of t f(t) exists, and has the same abscissa of convergence as F(p). By definition, we have

$$F(p) = \int_0^\infty f(t)\, e^{-pt}\, dt$$

and, from Leibnitz's rule for differentiating under the integral sign:

$$\frac{d\,F(p)}{dp} = -\int_0^\infty t\,f(t)\, e^{-pt}\, dt$$

which proves rule (15.25). The general rule is easily shown:

$$t^n f(t) \to (-1)^n \frac{d^n F(p)}{dp^n} \tag{15.26}$$

15.3.6 Initial-value theorem

Starting with rule (15.13) for the derivative:

$$\frac{df(t)}{dt} \rightarrow pF(p) - f(0)$$

from which we find for the function pF(p):

$$pF(p) = \int_0^\infty \frac{df(t)}{dt}\, e^{-pt}\, dt + f(0)$$

As p approaches infinity, the above integral approaches zero, and pF(p) approaches f(0).
Thus, we arrive at the ***initial-value theorem***:

$$f(0) = \lim_{p \to \infty} pF(p) \qquad (15.27)$$

15.3.67 Laplace transform dictionary

To facilitate the use of the transforms obtained, we have grouped them in table 15.3.

Table 15.3

	$f(t)$	$F(p) = \int_0^\infty f(t)\, e^{-pt}\, dt$
1	$\lambda f(t) + \mu g(t)$	$\lambda F(p) + \mu G(p)$
2	$\delta(t)$ (Dirac pulse)	1
3	$u(t)$ (unit step)	$1/p$
4	$e^{-\alpha t}$	$1/(p+\alpha)$
5	$\sin \omega t$	$\omega/(p^2+\omega^2)$
6	$\cos \omega t$	$p/(p^2+\omega^2)$
7	$\dfrac{e^{\alpha t} \sin \omega t}{\omega}$	$\dfrac{1}{(p-\alpha)^2+\omega^2}$
8	$e^{\alpha t} \cos \omega t$	$\dfrac{p-\alpha}{(p-\alpha)^2+\omega^2}$
9	$t^n/n!$	$1/p^{n+1}$
10	$e^{-\alpha t} f(t)$	$F(p+\alpha)$
11	$u(t-\tau) f(t-\tau) d\tau$	$e^{-p\tau} F(p)$
12	$\int_0^t f(\tau) g(t-\tau)\, d\tau$	$F(p)\, G(p)$
13	$\int_0^t f(\tau)\, d\tau$	$F(p)/p$
14	$t^n f(t)$	$(-1)^n \dfrac{d^n F(p)}{dp^n}$
15	$f'(t)$	$pF(p) - f(0)$

Table 15.3 (cont'd)

16	$f''(t)$	$p^2 F(p) - p f(0) - f'(0)$
17	$f^{(n)}(t)$	$p^n F(p) - p^{n-1} f(0) - \ldots - f^{(n-1)}(0)$
18	$\lim_{t \to \infty} f(t)$	$\lim_{p \to 0} p F(p)$
19	$f(0)$	$\lim_{p \to \infty} p F(p)$
20	$t e^{-\alpha t}$	$1/(p + \alpha)^2$
21	$e^{-\alpha t} \sin \omega t$	$\omega/[(p + \alpha)^2 + \omega^2]$
22	$e^{-\alpha t} \cos \omega t$	$(p + \alpha)/[(p + \alpha)^2 + \omega^2]$
23	$\text{sh}\, \omega t$	$\omega/(p^2 - \omega^2)$
24	$\text{ch}\, \omega t$	$p/(p^2 - \omega^2)$

15.4 SOLUTION OF DIFFERENTIAL EQUATIONS

The Laplace transformation is a powerful method for solving ordinary ***linear*** differential equations with ***constant coefficients.***

Given an *n*th order linear differential equation with constant coefficients:

$$a_0 \frac{d^n x}{dt^n} + a_1 \frac{d^{n-1} x}{dt^{n-1}} + \ldots + a_{n-1} \frac{dx}{dt} + a_n x(t) = y(t) \tag{15.28}$$

we wish to find the solution x(t) of this equation for $t \geq 0$, satisfying the n initial conditions:

$$x(0) = x_0, \quad \dot{x}(0) = \dot{x}_0, \ldots, x^{(n-1)}(0) = x_0^{(n-1)}$$

The Laplace transformation reduces this problem to the solution of an algebraic equation. Another advantage is that it takes care of the initial conditions so that the determination of a general solution is avoided by solving the corresponding homogeneous equation and adding a particular solution of the problem.

To illustrate this approach, let us assume that the transforms of x(t) and y(t) are X(p) and Y(p), respectively. Multiplying the two members of the equality (15.28) by exp (–pt), and integrating between the limits 0 and ∞ with respect to t, we have, using the property of linearity (15.5) and the rule of the derivative (15.13):

$$\begin{aligned} &a_0 [p^n X(p) - p^{n-1} x(0) - p^{n-2} \dot{x}(0) - \ldots - x^{(n-1)}(0)] \\ &\quad + a_1 [p^{n-1} X(p) - p^{n-2} x(0) - p^{n-3} \dot{x}(0) - \ldots - x^{(n-2)}(0)] \\ &\quad + \ldots + a_{n-1} [p X(p) - x(0)] + a_n X(p) = Y(p) \end{aligned} \tag{15.29}$$

Equation (15.29) is called the ***subsidiary equation*** of the given differential equation (15.28). Its solution is clearly:

$$X(p) = \frac{Y(p)}{a_0 p^n + a_1 p^{n-1} + \ldots + a_{n-1} p + a_n} + \frac{H(p)}{a_0 p^n + a_1 p^{n-1} + \ldots + a_{n-1} p + a_n} \tag{15.30}$$

where all the initial conditions have been incorporated in the polynomial H(p) of degree $n-1$; the coefficients of H(p), therefore, are known.

If the transform Y(p) of the excitation y(t) is a rational function, X(p) will also be a rational function, since H(p) is a polynomial. This shows that in many cases of practical importance, the solution of the subsidiary equation is a rational function. Because each ***rational function*** can be expressed in terms of ***partial fractions*** the inverse transform is easily obtained by using table 15.3.

15.5 LINEAR SYSTEMS

Let us again consider the differential equation with constant coefficients:

$$a_0 \frac{d^n x}{dt^n} + a_1 \frac{d^{n-1} x}{dt^{n-1}} + \ldots + a_{n-1} \frac{dx}{dt} + a_n x(t) = y(t) \tag{15.31}$$

satisfying the initial conditions:

$$x(0) = x_0, \ \dot{x}(0) = \dot{x}_0, \ldots, x^{(n-1)}(0) = x_0^{(n-1)} \tag{15.32}$$

A physical realization of this differential equation is a particular case of the class of ***stationary linear filters*** whose schematic is shown in figure 15.4.

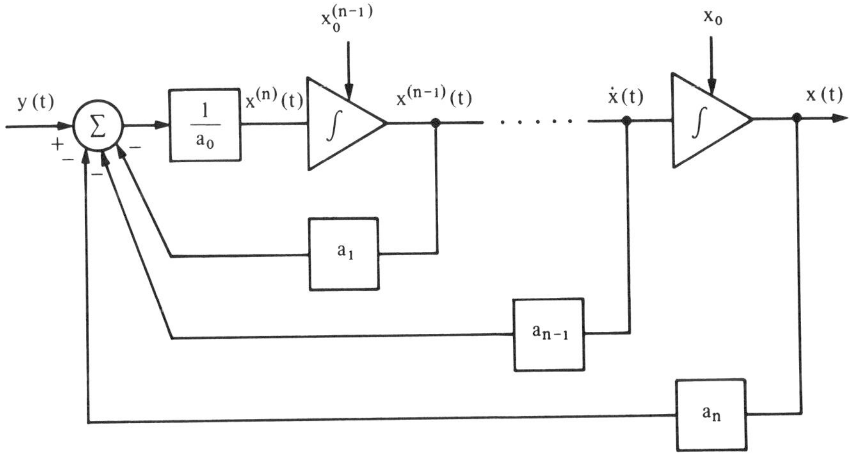

Fig. 15.4

Applying the Laplace transform to the left and right sides of equation (15.31), we have just found for the image X(p):

$$X(p) = \frac{Y(p)}{a_0 p^n + a_1 p^{n-1} + \ldots + a_n} + \frac{H(p)}{a_0 p^n + a_1 p^{n-1} + \ldots + a_n} \tag{15.33}$$

The polynomial in the denominator of the right-hand terms of relation (15.33) is the ***characteristic polynomial*** of the differential equation (15.31). To study the behavior of the linear system described by equation (15.31), we first consider the zero-input case, i.e., we set $y(t) = 0$, which implies that $Y(p) = 0$. We therefore have for (15.33):

$$X(p) = \frac{H(p)}{a_0 p^n + a_1 p^{n-1} + \ldots + a_n} \tag{15.34}$$

By expressing this rational function in terms of partial fractions, we establish that a ***pole*** p_i of order k gives the following contribution $X_i(p)$ to the transform X(p):

$$\frac{c_0}{p - p_i} + \frac{c_1}{(p - p_i)^2} + \ldots + \frac{c_{k-1}}{(p - p_i)^k} = X_i(p) \tag{15.35}$$

for which the inverse tranform $x_i(t)$ will be

$$X_i(p) \rightarrow x_i(t) = \left(c_0 + c_1 t + \ldots + c_{k-1} \frac{t^{k-1}}{(k-1)!} \right) e^{p_i t} \tag{15.36}$$

In all cases of practical importance, the initial conditions (15.32) are not to intervene in the zero-input response. Consequently, ***all of the poles*** p_i which are roots of the characteristic equation of the differential equation (15.31), must be located ***in the left half-plane*** Re $p < 0$ ***so that the zero input response decreases exponentially.*** Such a system is called ***exponentially stable***. Thus, the initial conditions will no longer intervene after a certain time in an exponentially stable system, which leads to setting them to zero in equation (14.33). Thus, H(p) vanishes identically, and we find for the ratio of the response transform X(p) and the excitation transform Y(p):

$$\frac{X(p)}{Y(p)} = \frac{1}{a_0 p^n + a_1 p^{n-1} + \ldots + a_n} = F(p) \tag{15.37}$$

The function F(p) thus defined is called the ***transfer function*** of the system, and plays an important role in the analysis of the behavior of a system with sinusoidal excitation. To this end, we shall examine the steady-state response of an exponentially stable system to the input signal $y(t) = \cos \omega t$. In passing from the function $\cos \omega t$ to exponentials, we first study the response of the system to the excitation $y_1(t) = \exp(j \omega t)$ whose transform equals $Y_1(p) = 1/(p - j\omega)$. Since we can neglect the contribution due to the initial conditions in (15.33), we have

$$X(p) = \frac{1}{a_0 p^n + a_1 p^{n-1} + \ldots + a_n} \cdot \frac{1}{p - j\omega} = F(p) \frac{1}{p - j\omega} \tag{15.38}$$

By expanding this rational function into partial fractions, we obtain

$$X(p) = \sum_{k=1}^{n} \frac{c_k}{p - p_k} + \frac{c_{n+1}}{p - j\omega} \tag{15.39}$$

where in order to clarify the considerations, it has been assumed that all of the poles p_k are simple poles. Each term $c_k/(p-p_k)$ has as its inverse transform $c_k \exp(p_k t)$. Because the system is assumed to be exponentially stable, all the poles are located in the left half-plane $\operatorname{Re} p < 0$ and consequently the ***transient response*** consisting of the sum of the exponentials $c_k \exp(p_k t)$, $k = 1, 2, \ldots, n$, approaches zero as $t \to \infty$. The ***steady-state response***, finally, is the inverse transform of the term $c_{n+1}/(p - j\omega)$, i.e.,

$$x_1(t) = c_{n+1} e^{j\omega t} \tag{15.40}$$

The coefficient c_{n+1} is given by the limit (see (15.38) and (15.39)):

$$c_{n+1} = \lim_{p \to j\omega} X(p)(p - j\omega) = F(j\omega) \tag{15.41}$$

from which we obtain for the steady-state response $x_1(t)$ to the excitation $y_1(t) = \exp(j\omega t)$:

$$x_1(t) = F(j\omega) e^{j\omega t} \tag{15.42}$$

In an analogous fashion, we obtain the steady-state response $x_2(t)$ to the excitation $y_2(t) = \exp(-j\omega t)$:

$$x_2(t) = F(-j\omega t) e^{-j\omega t} \tag{15.43}$$

Thus, the steady-state response $x(t)$ to the sinusoidal excitation $\cos\omega t$ can be expressed in the form of the following sum:

$$x(t) = \frac{1}{2}(x_1(t) + x_2(t)) = \frac{F(j\omega)e^{j\omega t} + F(-j\omega)e^{-j\omega t}}{2} \tag{15.44}$$

Because the coefficients a_k of the differential equaition (15.27) are real numbers, we have

$$F(-j\omega) = \overline{F}(j\omega) = Re^{\,j\varphi}$$

where we have set

$$F(j\omega) = Re^{j\varphi}$$

Relation (15.44) is thus written:

$$x(t) = R\,\frac{e^{j(\omega t + \varphi)} + e^{-j(\omega t + \varphi)}}{2} = R\cos(\omega t + \varphi)$$

or, using the transfer function $F(p)$:

$$x(t) = |F(j\omega)| \cos(\omega t + \arg F(j\omega)) \tag{15.45}$$

The modulus $|F(j\omega)|$ is called the ***amplitude response***, and the argument arg $F(j\omega)$ the ***phase response*** of the system.

15.6 APPLICATIONS OF THE LAPLACE TRANSFORM

15.6.1 Example: Solution of a differential equation

Find the solution of the homogeneous differential equation:

$$\frac{d^3x}{dt^3} + x(t) = 0$$

satisfying the initial conditions $x_0 = 1$, $\dot{x}_0 = 3$, $\ddot{x}_0 = 8$ for $t = 0$.
Let us designate the Laplace transform of x(t) by X(p); then (table 15.3):

$$p^3X(p) - p^2 - 3p - 8 + X(p) = 0$$

from which we get

$$X(p) = \frac{p^2 + 3p + 8}{p^3 + 1} = \frac{p^2 + 3p + 8}{(p+1)(p^2 - p + 1)}$$

The expansion of the rational function into partial fractions is the following:

$$X(p) = \frac{2}{p+1} + \frac{6-p}{p^2 - p + 1}$$

$$= \frac{2}{p+1} - \frac{p - \frac{1}{2}}{\left(p - \frac{1}{2}\right)^2 + \left(\sqrt{\frac{3}{4}}\right)^2} + \frac{11}{\sqrt{3}} \frac{\sqrt{\frac{3}{4}}}{\left(p - \frac{1}{2}\right)^2 + \left(\sqrt{\frac{3}{4}}\right)^2}$$

from which we obtain from table 15.3 for the solution x(t):

$$x(t) = 2e^{-t} + e^{t/2}\left(-\cos\sqrt{\frac{3}{4}}\,t + \frac{11}{\sqrt{3}}\sin\sqrt{\frac{3}{4}}\,t\right)$$

15.6.2 Example: Application of the convolution theorem

To illustrate the utility of the convolution theorem, we shall apply it to the following problem. Find the solution of the nonhomogeneous differential equation:

$$\frac{d^2x}{dt^2} + x(t) = y(t)$$

satisfying the initial conditions $x_0 = \dot{x}_0 = 0$ for $t = 0$.

If X(p) and Y(p) designate the transforms of x(t) and y(t), respectively, we obtain for the subsidiary equation:

$$p^2X(p) + X(p) = Y(p)$$

from which we get

$$X(p) = \frac{1}{p^2+1}\,Y(p)$$

Because

$$\frac{1}{p^2+1} \rightarrow \sin t$$

and

$$Y(p) \rightarrow y(t)$$

we find, according to the convolution theorem (15.23), for the solution x(t):

$$x(t) = \int_0^t y(\tau)\sin(t-\tau)\,d\tau$$

15.6.3 Example: Steady-state response of a RCL circuit

Consider an electrical circuit, composed of an induction coil L, a resistor R, and a capacitor C in series, to which a voltage v(t) is applied (fig. 15.5).

We designate the current in the circuit by i(t). Thus, Kirchhoff's voltage law yields the following equation:

$$v(t) = Ri + L\,\frac{di}{dt} + \frac{1}{C}\left\{\int_0^t i(\tau)\,d\tau + Q_0\right\} \tag{15.46}$$

where Q_0 is the initial charge of the capacitor. Denoting the transforms of v(t) and i(t) by V(p) and I(p), respectively, we obtain for the subsidiary equation corresponding to (15.46)

$$V(p) = RI(p) + L[pI(p) - i(0)] + \frac{1}{C}\left\{\frac{I(p)}{p} + \frac{Q_0}{p}\right\}$$

where i(0) is the initial current across the inductance. Hence, we solve the above equation for I(p):

$$I(p) = \frac{p}{Lp^2 + Rp + 1/C} V(p) - \frac{Q_0/C - pLi(0)}{Lp^2 + Rp + 1/C}$$

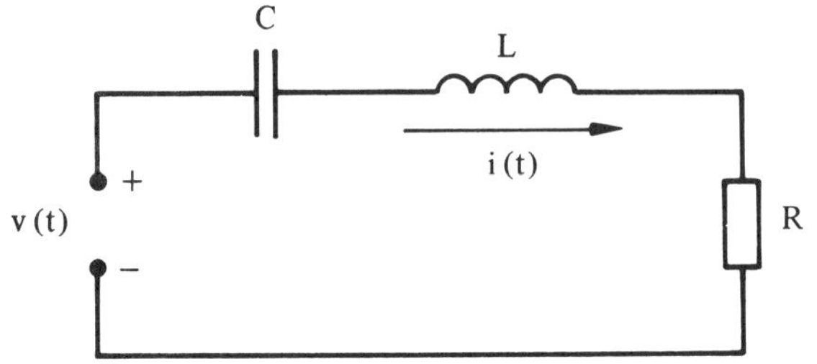

Fig. 15.5

Because the roots of the characteristic equation:

$$Lp^2 + Rp + 1/C = 0$$

are located in the left half-plane (we have $R > 0$, $L > 0$, and $C > 0$), the circuit is exponentially stable, and we find the transfer function F(p) by cancelling the initial conditions and determining the ratio of the response transform I(p) and the excitation transform V(p):

$$F(p) = \frac{I(p)}{V(p)} = \frac{p}{Lp^2 + Rp + 1/C}$$

We arrive at the frequency response by setting $p = j\omega$, from which we obtain for the amplitude response of the circuit, squared:

$$|F(j\omega)|^2 = \frac{\omega^2}{(1/C - \omega^2 L)^2 + \omega^2 R^2}$$

and we obtain for the phase response:

$$\arg F(j\omega) = \arctan[(1/C - \omega^2 L)/R\omega]$$

Finally, the steady-state response to the excitation:

$$v(t) = v_0 \cos \omega t$$

is given by (section 15.5):

$$i(t) = v_0 \, | F(j\omega) | \cos(\omega t + \arg F(j\omega))$$

15.6.4 Example: Tracking circuit

We shall consider the tracking circuit shown in figure 15.6.

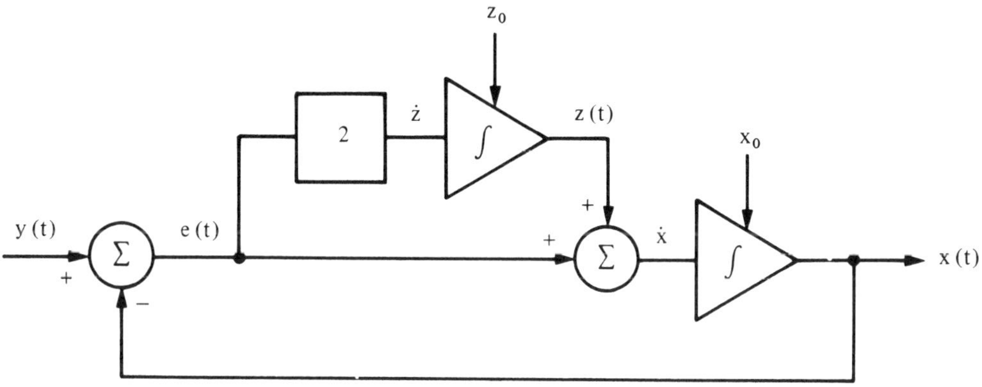

Fig. 15.6

The circuit is described by the two differential equations

$$\dot{x} = z + (y - x)$$

$$\dot{z} = 2(y - x)$$

subject to the initial conditions $x(0) = x_0$ and $z(0) = z_0$.
For the subsidiary equations, we obtain

$$pX(p) - x_0 = Z(p) + [Y(p) - X(p)]$$

$$pZ(p) - z_0 = 2[Y(p) - X(p)]$$

and eliminating Z(p), we have for X(p):

$$X(p) = \frac{p+2}{p^2+p+2}\,Y(p) + \frac{px_0 + z_0}{p^2+p+2}$$

The roots of the characteristic equation $p^2 + p + 2 = 0$ are

$$p_{1,2} = -\frac{1}{2} \pm j\,\frac{\sqrt{7}}{2}$$

which are located in the left half-plane Re $p<0$. The circuit considered is, therefore, exponentially stable, and canceling the initial conditions x_0 and z_0, we find for the transfer function F(p):

$$F(p) = \frac{X(p)}{Y(p)} = \frac{p+2}{p^2+p+2}$$

or

$$X(p) = \frac{p+2}{p^2+p+2}\,Y(p)$$

Let us now consider the ***tracking error*** $e(t) = y(t) - x(t)$ as $t \to \infty$. For its transform, we obtain

$$E(p) = Y(p) - X(p) = \frac{p^2}{p^2 + p + 2} Y(p)$$

If, for example,

$$y(t) = \frac{t^n}{n!}$$

then,

$$Y(p) = \frac{1}{p^{n+1}}$$

and for the final tracking error (see final-value theorem, section 15.3.4):

$$\lim_{t \to \infty} e(t) = \lim_{p \to 0} pE(p) = \lim_{p \to 0} \frac{p^3}{p^{n+1}(p^2 + p + 2)}$$

$$= \begin{cases} 0 & \text{for } n = 0 \\ 0 & \text{for } n = 1 \\ 1/2 & \text{for } n = 2 \\ \infty & \text{for } n \geq 3 \end{cases}$$

Therefore, the tracking circuit (fig. 15.6) is capable of tracking a uniform movement without final tracking error ($n = 1$). Conversely, it develops a tracking error equal to $1/2$ when tracking a movement with constant acceleration equal to one ($n = 2$), and can no longer follow a nonconstant acceleration ($n \geq 3$). This is why the circuit under consideration is a ***second-order tracking circuit***.

15.6.5 Example: Stability of nonlinear systems

Let us consider the system of two first-order nonlinear differential equations:

$$\begin{aligned} \dot{x}(t) &= F(x,y) \\ \dot{y}(t) &= G(x,y) \end{aligned} \tag{15.47}$$

where $x(t)$ and $y(t)$ are to be determined.

In general, such a system cannot be explicitly integrated, but its behavior can be easily studied in the neighborhood of an equilibrium point. An ***equilibrium point*** or ***stationary point*** (x_0, y_0) of system (15.47) is a point such that

$$F(x_0,y_0) = G(x_0,y_0) = 0 \tag{15.48}$$

In the neighborhood of an equilibrium point (x_0, y_0), we write

$$x(t) = x_0 + \Delta x(t)$$
$$y(t) = y_0 + \Delta y(t)$$

and substituting these expressions into the system of differential equations (15.47), we have

$$\begin{aligned} \Delta \dot{x} &= F(x_0 + \Delta x, y_0 + \Delta y) \\ \Delta \dot{y} &= G(x_0 + \Delta x, y_0 + \Delta y) \end{aligned} \tag{15.49}$$

By expanding the two functions F and G at the point (x_0, y_0) in Taylor series and retaining only the linear terms, we obtain for small Δx and Δy:

$$\begin{aligned} \Delta \dot{x} &= F(x_0,y_0) + \frac{\partial F}{\partial x} \Delta x + \frac{\partial F}{\partial y} \Delta y \\ \Delta \dot{y} &= G(x_0,y_0) + \frac{\partial G}{\partial x} \Delta x + \frac{\partial G}{\partial y} \Delta y \end{aligned} \tag{15.50}$$

Because (x_0, y_0) is an equilibrium point (15.48), it follows, in matrix notation:

$$\begin{bmatrix} \Delta \dot{x} \\ \Delta \dot{y} \end{bmatrix} = \begin{bmatrix} \dfrac{\partial F}{\partial x} & \dfrac{\partial F}{\partial y} \\ \dfrac{\partial G}{\partial x} & \dfrac{\partial G}{\partial y} \end{bmatrix}_{(x_0,y_0)} \begin{bmatrix} \Delta x \\ \Delta y \end{bmatrix} \tag{15.51}$$

which constitutes a system of linear differential equations with constant coefficients for the terms Δx and Δy.

Let us designate by $\Delta X(p)$ and $\Delta Y(p)$ the transforms of $\Delta x(t)$ and $\Delta y(t)$, respectively. Thus, we obtain for the subsidiary matrix equation:

$$p \begin{bmatrix} \Delta X(p) \\ \Delta Y(p) \end{bmatrix} - \begin{bmatrix} \Delta x(0) \\ \Delta y(0) \end{bmatrix} = \begin{bmatrix} \dfrac{\partial F}{\partial x} & \dfrac{\partial F}{\partial y} \\ \dfrac{\partial G}{\partial x} & \dfrac{\partial G}{\partial y} \end{bmatrix}_{(x_0,y_0)} \begin{bmatrix} \Delta X(p) \\ \Delta Y(p) \end{bmatrix} \tag{15.52}$$

and solving this equation for the unknown vector $[\Delta X(p), \Delta Y(p)]$:

$$\begin{bmatrix} \Delta X(p) \\ \\ \Delta Y(p) \end{bmatrix} = \begin{bmatrix} p - \dfrac{\partial F}{\partial x} & -\dfrac{\partial F}{\partial y} \\ \\ -\dfrac{\partial G}{\partial x} & p - \dfrac{\partial G}{\partial y} \end{bmatrix}^{-1}_{(x_0, y_0)} \begin{bmatrix} \Delta x_0 \\ \\ \Delta y_0 \end{bmatrix} \tag{15.53}$$

If the eigenvalues of system (15.53), i.e., the roots of the determinant:

$$\begin{vmatrix} p - \dfrac{\partial F}{\partial x} & -\dfrac{\partial F}{\partial y} \\ \\ -\dfrac{\partial G}{\partial x} & p - \dfrac{\partial G}{\partial y} \end{vmatrix}_{(x_0, y_0)}$$

are located in the left half-plane Re $p<0$, the trajectories in the xy phase plane satisfying the differential equations (15.47) approach the equilibrium point (x_0, y_0) as $t \to \infty$. In this case, the system (15.47) is said to be ***stable*** at the point (x_0, y_0). It is sufficient that a single eigenvalue of the system (15.53) is located in the right half-plane Re $p>0$ for system (15.47) to become ***unstable***, i.e., the terms $\Delta x(t)$ and $\Delta y(t)$ increase infinitely as $t \to \infty$.

As an example, let us study the movement of the ***simple pendulum*** described by the second-order differential equation:

$$\ddot{\varphi} + \frac{g}{l} \sin \varphi = 0 \tag{15.54}$$

If we set $g/l = 1$, and if we introduce the phase coordinates:

$$x(t) = \varphi(t)$$
$$y(t) = \dot{\varphi}(t)$$

then equation (15.54) corresponds to the following system of first-order differential equations:

$$\begin{aligned} \dot{x} &= y = F(x,y) \\ \dot{y} &= -\sin x = G(x,y) \end{aligned} \tag{15.55}$$

The equilibrium points (x_0, y_0) of the system are determined by the conditions:

$$\dot{x} = \dot{y} = 0$$

from which we get

$$x_0 = k\pi, \; y_0 = 0; \quad k = 0,1,2,\ldots$$

For $k=0$ ($x_0=0$, $y_0=0$), we obtain for the matrix of the system:

$$\begin{bmatrix} \frac{\partial F}{\partial x} & \frac{\partial F}{\partial y} \\ \frac{\partial G}{\partial x} & \frac{\partial G}{\partial y} \end{bmatrix}_{(0,0)} = \begin{bmatrix} 0 & 1 \\ -1 & 0 \end{bmatrix}$$

and equation (15.53) becomes

$$\begin{bmatrix} \Delta X(p) \\ \Delta Y(p) \end{bmatrix} = \begin{bmatrix} p & -1 \\ 1 & p \end{bmatrix}^{-1} \begin{bmatrix} \Delta x_0 \\ \Delta y_0 \end{bmatrix} = \frac{1}{p^2+1} \begin{bmatrix} p\Delta x_0 + \Delta y_0 \\ -\Delta x_0 + p\Delta y_0 \end{bmatrix}$$

According to table 15.3, we find for the inverse transform:

$$\Delta x(t) = \Delta x_0 \cos t + \Delta y_0 \sin t$$
$$\Delta y(t) = -\Delta x_0 \sin t + \Delta y_0 \cos t$$

Because

$$[\Delta x_0 \Delta x(t) + \Delta y_0 \Delta y(t)]^2 + [\Delta y_0 \Delta x(t) - \Delta x_0 \Delta y(t)]^2 = [\Delta x_0^2 + \Delta y_0^2]^2$$

the trajectory of system (15.55) in the phase plane (x,y) is an ellipse centered at the origin. Because the trajectory neither approaches nor moves away from the origin as $t \to \infty$, this type of equilibrium point, which is neither stable nor unstable, is called a ***center***. The eigenvalues of the system, which are the poles of the rational fraction $1/(p^2+1)$, are $p_{1,2} = \pm j$. They are located on the imaginary axis $\text{Re } p = 0$, which explains the behavior of the system at the center.

For $k=1$ ($x_0=\pi$, $y_0=0$), the matrix of the system equals

$$\begin{bmatrix} \frac{\partial F}{\partial x} & \frac{\partial F}{\partial y} \\ \frac{\partial G}{\partial x} & \frac{\partial G}{\partial y} \end{bmatrix}_{(\pi,0)} = \begin{bmatrix} 0 & 1 \\ 1 & 0 \end{bmatrix}$$

from which we obtain for equation (15.53):

$$\begin{bmatrix} \Delta X(p) \\ \Delta Y(p) \end{bmatrix} = \begin{bmatrix} p & -1 \\ -1 & p \end{bmatrix}^{-1} \begin{bmatrix} \Delta x_0 \\ \Delta y_0 \end{bmatrix} = \frac{1}{p^2-1} \begin{bmatrix} p\Delta x_0 + \Delta y_0 \\ \Delta x_0 + p\Delta y_0 \end{bmatrix}$$

According to table 15.3, we find for the inverse transform

$$\Delta x(t) = \Delta x_0 \mathrm{ch}\, t + \Delta y_0 \,\mathrm{sh}\, t$$
$$\Delta y(t) = \Delta x_0 \mathrm{sh}\, t + \Delta y_0 \,\mathrm{ch}\, t$$

and since

$$[\Delta x_0 \Delta x(t) - \Delta y_0 \Delta y(t)]^2 - [\Delta y_0 \Delta x(t) - \Delta x_0 \Delta y(t)]^2 = [\Delta x_0^2 - \Delta y_0^2]^2$$

hence, the trajectory of the system (15.55) in the phase plane (x,y) is a hyperbola. The equilibrium point $(\pi,0)$ is, therefore, unstable as the trajectory moves away from it as $t \to \infty$. The eigenvalues of the system, which are the poles of the rational fraction $1/(p^2-1)$, are $p_{1,2} = \pm 1$. One of them is located in the right half-plane Re $p > 0$, which explains why the system is unstable at the point $(\pi,0)$.

Chapter 16

Introduction to the Calculus of Variations

16.1 INTRODUCTION

An important class of problems involves the determination of a function subject to certain conditions so as to maximize or minimize a certain definite integral. For example, let us consider a material particle P, which, affected only by the force of gravity, slides without friction along an inclined plane curve C, from point $P_0(0,0)$ to point $P_1(x_1, y_1)$. It is assumed that the y-axis is oriented vertically toward the bottom, i.e., in the direction of the action of the force (fig. 16.1).

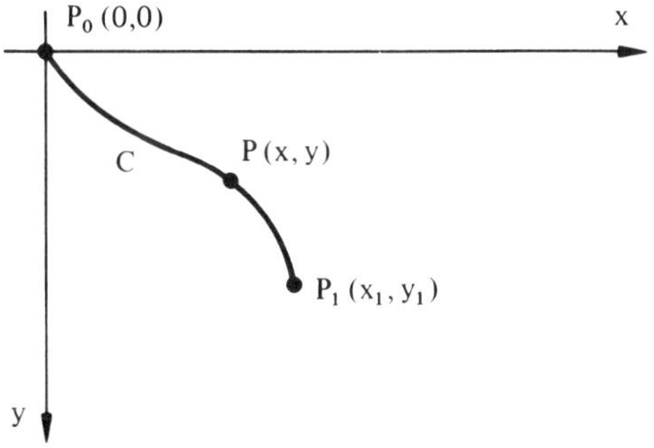

Fig. 16.1

The speed of P at any instant can be determined by using the principle of energy conservation:

$$\frac{m}{2} v^2 = mgy$$

from which we obtain, introducing the arc length s as a parameter:

$$v = \frac{ds}{dt} = \sqrt{2gy}$$

The total time of descent along the curve C is given by

$$T = \int_C dt = \int_C \frac{dt}{ds}\, ds = \int_C \frac{1}{v}\, ds \qquad (16.1)$$

with

$$ds = \sqrt{1 + y'^2}\, dx$$

thus,

$$T = \int_{x_0=0}^{x_1} \frac{\sqrt{1+y'^2}}{\sqrt{2gy}}\, dx$$

How should the curve C be chosen so that the descent occurs in the shortest possible time? Or, stated otherwise, how to determine the function y(x) subject to the conditions $y(0) = 0$ and $y(x_1) = y_1$ such that the integral (16.1) is a minimum? This problem is known as the brachistochrone problem (*brachistos* = shortest, *chronos* = time).

16.2 EULER'S EQUATION FOR THE SIMPLE CASE

In the general case, we want to find the function y(x) subject to the end conditions $y(x_0) = y_0$ and $y(x_1) = y_1$, such that, for some given function $F(x,y,y')$, the definite integral:

$$I = \int_{x_0}^{x_1} F(x, y(x), y'(x))\, dx \qquad (16.2)$$

is either a maximum or minimum, also called an extremum.

We suppose that $F(x,y,y')$ has continuous second-order derivatives with respect to its three arguments and we require that the unknown function y(x) has two derivatives in the considered interval.

To fix ideas, let us deduce a ***necessary condition***, which the function y(x) must satisfy, so that the integral (16.2) is a minimum. To this end, we consider a function $\eta(x)$, which vanishes at the end points of the integration interval:

$$\eta(x_0) = \eta(x_1) = 0 \qquad (16.3)$$

Then, we construct the function $\widetilde{y}(x)$:

$$\widetilde{y}(x) = y(x) + \epsilon\eta(x)$$

where ϵ is a parameter. This new function satisfies the same conditions at the end points as the function y(x). Substituting this modified function into the integral (16.2), I becomes a function of the parameter ϵ:

$$I(\epsilon)=\int_{x_0}^{x_1} F(x,y(x)+\epsilon\eta(x),y'(x)+\epsilon\eta'(x))\,dx \tag{16.4}$$

Because $I(\epsilon)$ takes on its minimum value when $\epsilon=0$, its derivative is zero for $\epsilon=0$. Differentiating under the integral sign of (16.4), we obtain for the derivative

$$I'(0)=0=\int_{x_0}^{x_1}\left[\frac{\partial F(x,y,y')}{\partial y}\eta(x)+\frac{\partial F(x,y,y')}{\partial y'}\eta'(x)\right]dx \tag{16.5}$$

Integration by parts of the second term on the right side of (16.5) gives

$$I'(0)=\frac{\partial F}{\partial y'}\eta(x)\Bigg|_{x_0}^{x_1}+\int_{x_0}^{x_1}\eta(x)\left[\frac{\partial F}{\partial y}-\frac{d}{dx}\left(\frac{\partial F}{\partial y'}\right)\right]dx=0 \tag{16.6}$$

The term outside the integral sign vanishes since $\eta(x_0)=\eta(x_1)=0$. Further, since $\eta(x)$ is an arbitrary function, the coefficient of $\eta(x)$ under the integral sign must also vanish identically. To show this, we proceed in an indirect fashion: we assume that this coefficient does not vanish identically. Then, we choose the function $\eta(x)$ to be positive wherever

$$\frac{\partial F}{\partial y}-\frac{d}{dx}\left(\frac{\partial F}{\partial y'}\right)>0$$

and negative where this same expression is negative. Thus, the function to be integrated in (16.6) is positive, which is in contradiction to the fact that the integral is zero. Therefore, we have the following necessary condition which the solution must satisfy:

$$\frac{\partial F}{\partial y}-\frac{d}{dx}\left(\frac{\partial F}{\partial y'}\right)=0 \tag{16.7}$$

This is the so-called ***Euler equation***.

As an example, in the case of the brachistochrone (16.1), we have

$$F(x,y,y')=\frac{\sqrt{1+y'^2}}{\sqrt{2gy}} \tag{16.8}$$

which does not depend explicitly on x, and we can easily deduce the first integral of Euler's equation (16.7):

$$F - y' \frac{\partial F}{\partial y'} = c \tag{16.9}$$

where c is a constant. In effect, we have

$$\frac{dF}{dx} = \frac{\partial F}{\partial x} + \frac{\partial F}{\partial y} y' + \frac{\partial F}{\partial y'} y'' \tag{16.10}$$

and

$$\frac{d}{dx}\left(y' \frac{\partial F}{\partial y'}\right) = y' \frac{d}{dx}\left(\frac{\partial F}{\partial y'}\right) + \frac{\partial F}{\partial y'} y'' \tag{16.11}$$

Subtracting (16.11) from (16.10), we obtain

$$\frac{d}{dx}\left(F - y' \frac{\partial F}{\partial y'}\right) = \frac{\partial F}{\partial x} + y' \left[\frac{\partial F}{\partial y} - \frac{d}{dx}\left(\frac{\partial F}{\partial y'}\right)\right] \tag{16.12}$$

Because F does not depend explicitly on x, we have $\partial F / \partial x = 0$, and since the term in brackets on the right side of (16.12) is zero, it follows that

$$\frac{d}{dx}\left(F - y' \frac{\partial F}{\partial y'}\right) = 0$$

from which we obtain, after integration:

$$F - y' \frac{\partial F}{\partial y'} = c$$

For the brachistochrone problem under consideration, the preceding relation is equivalent to the following:

$$\sqrt{y}\,\sqrt{1 + y'^2} = \frac{1}{c\sqrt{2g}} = c_1$$

and in solving for y′, we find

$$y' = \sqrt{\frac{a - y}{y}}$$

with $\sqrt{a} = c_1$. Separating the variables and integrating a second time, we have

$$\int dx = \int \sqrt{\frac{y}{a - y}}\, dy$$

and making the substitution:

$$y = a \sin^2 \varphi = \frac{a}{2}(1 - \cos 2\varphi)$$

we get

$$x = 2a\int \sin^2\varphi\, d\varphi + c_2$$
$$= a\int (1 - \cos 2\varphi) d\varphi + c_2 = \frac{a}{2}(2\varphi - \sin 2\varphi) + c_3$$

Thus, the parametric equations of the curve sought are

$$x = b(\phi - \sin\phi) + c_3$$
$$y = b(1 - \cos\phi)$$

where we have set $b = a/2$ and $\phi = 2\varphi$, and c_3 is zero as the curve passes through the origin.

The resulting curve is therefore a cycloid, which is the path of a fixed point on a circle of radius b as it rolls along the x axis (fig. 16.2). The radius b is determined by specifying the end point P_1 (x_1, y_1).

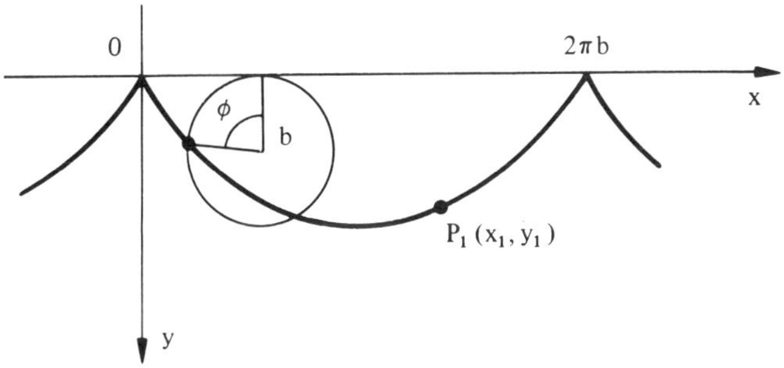

Fig. 16.2

Let us again emphasize that Euler's equation is a necessary condition. The formulation of *sufficient* conditions is much more complicated, but in the majority of cases they are not needed. If the solution to Euler's equation has been found, we may prove, often by using physical considerations, that this solution provides the extremum sought.

16.3 ISOPERIMETRIC PROBLEMS

If we consider the problem of finding the function y(x), which makes a given integral:

$$I_1 = \int_{x_0}^{x_1} F(x,y,y')\,dx \tag{16.13}$$

a maximum or minimum, but at the same time keeps the integral:

$$I_2 = \int_{x_0}^{x_1} G(x,y,y')\,dx \tag{16.14}$$

equal to some constant, we speak of an ***isoperimetric problem***. The functions F and G are given functions of x,y, and y′.

A typical example of this type of problem is the following problem: to find, from among all the curves of given length l joining two points A and B, that which will limit the surface of greatest area with the line segment $\overline{AB}$ (fig. 16.3).

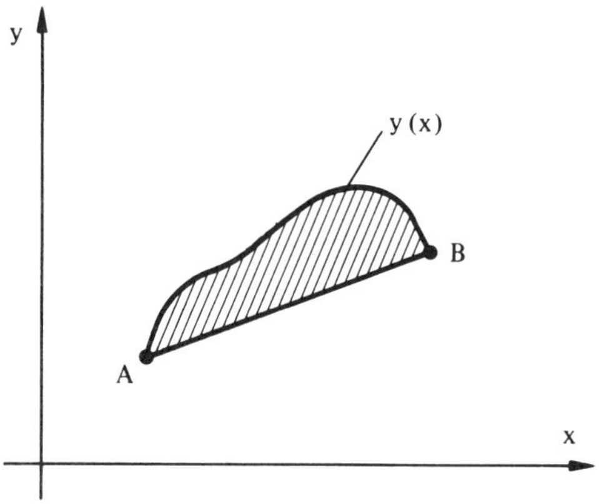

Fig. 16.3

This problem becomes the search for the maximum of the integral:

$$I_1 = \int_{x_0}^{x_1} y\,dx \tag{16.15}$$

undcr thc condition:

$$I_2 = \int_{x_0}^{x_1} \sqrt{1 + y'^2}\,dx = l \tag{16.16}$$

We can reduce this problem to an ordinary variational problem without constraints by using the method of ***Lagrange multipliers***. To this end, we proceed in the following way. In the general case we set,

$$H = F + \lambda G$$

where λ is a real parameter, and determine the function which minimizes the integral:

$$I = \int_{x_0}^{x_1} H(x,y,y')\,dx$$

from which we obtain Euler's equation (16.7):

$$\frac{\partial H}{\partial y} - \frac{d}{dx}\left(\frac{\partial H}{\partial y'}\right) = 0 \tag{16.17}$$

whose solution y depends on x and λ: $y = y(x,\lambda)$. The value of the parameter λ will be determined using the constraint (16.14) that the solution $y(x,\lambda)$ must satisfy. For our example ((16.15) and (16.16)), we have

$$H = y + \lambda\sqrt{1 + y'^2} \tag{16.18}$$

Because this function H does not explicitly depend on x, we find as the first integral of Euler's equation (16.17) (see (16.9)):

$$H - y'\frac{\partial H}{\partial y'} = c_1$$

Substituting (16.18) into the preceding relation, we obtain

$$y + \lambda\sqrt{1 + y'^2} - \frac{\lambda y'^2}{\sqrt{1 + y'^2}} = c_1$$

from which we get

$$y' = \frac{\sqrt{\lambda^2 - (c_1 - y)^2}}{c_1 - y}$$

or, also,

$$\frac{(c_1 - y)\,dy}{\sqrt{\lambda^2 - (c_1 - y)^2}} = dx$$

After integration, we arrive at

$$x = \sqrt{\lambda^2 - (c_1 - y)^2} + c_2$$

i.e., the resulting curves are circles of radius $|\lambda|$. The parameter λ and the integration constants c_1 and c_2 are determined in such a way that the circle passes through the two given points A and B, and its arc length is equal to *l*.

16.4 GENERALIZATION

Euler's equation is easily deduced for the case in which the integrand depends on several functions $y_1(x), y_2(x), \ldots, y_n(x)$:

$$I = \int_{x_0}^{x_1} F(x; y_1, y_2, \ldots, y_n; y_1', y_2', \ldots, y_n')\, dx \tag{16.19}$$

The necessary conditions for an extremum are expressed by the system of n second-order differential equations:

$$\frac{\partial F}{\partial y_k} - \frac{d}{dx}\left(\frac{\partial F}{\partial y_k'}\right) = 0, \quad k = 1, 2, \ldots, n \tag{16.20}$$

The preceding equations play a fundamental role in mechanics and mathematical physics.

If T designates the kinetic energy of a mechanical system and V is its potential energy, then, according to ***Hamilton's principle***, any displacement of the system is effected such that the integral:

$$I = \int_{t_0}^{t_1} (T - V)\, dt \tag{16.21}$$

is a minimum. The function $L = T - V$ is called the ***Lagrangian*** of the system, and the corresponding Euler equations (16.20):

$$\frac{\partial L}{\partial x_k} - \frac{d}{dt}\left(\frac{\partial L}{\partial \dot{x}_k}\right) = 0, \quad k = 1, 2, \ldots, n \tag{16.22}$$

are known as the ***Lagrange equations***, where $x_k(t)$, $k = 1, 2, \ldots, n$, are the coordinates of the system.

Part IV
Complex Variables

Chapter 17

Elementary Function of a Complex Variable

17.1 REVIEW OF COMPLEX NUMBERS

We call a ***complex number*** z any expression of the form:

$$z = a + jb$$

where a and b are real numbers, and j is the ***imaginary unit*** defined by the relation:

$$j^2 = -1$$

The numbers a and b are called the ***real part*** and the ***imaginary part***, respectively, of the complex number z, which are written:

$$a = \text{Re } z$$

$$b = \text{Im } z$$

If $a = 0$, $b \neq 0$, the complex number $0 + jb = jb$ is a ***pure imaginary*** number; if $b = 0$, we have a real number.

Two complex numbers $z = a + jb$ and $\overline{z} = a - jb$ are called ***complex conjugate***.

To a complex number $z = a + jb$, there corresponds, in the plane, the point with Cartesian coordinates (a, b). Conversely, to each point with coordinates (a, b) of the xy-plane, there corresponds the complex number $z = a + jb$. The xy-plane in which the complex numbers are represented geometrically in this fashion is called the complex plane or simply z-plane (fig. 17.1). This is why there is no distinction between the notion of a complex number and that of a point of the complex plane.

By designating the polar coordinates of the point $z = a + jb$ by r and φ, we obtain

$$a = r \cos \varphi$$

$$b = r \sin \varphi$$

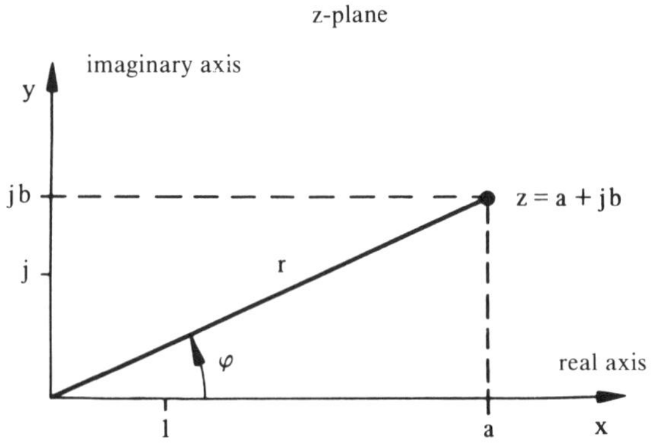

Fig. 17.1

The absolute value or ***modulus*** of z is denoted by $|z|$, and the ***argument*** of z is denoted by arg z. We have

$$|z| = r = \sqrt{a^2 + b^2}$$

$$\arg z = \varphi = \arctan \frac{b}{a}$$

From the relations:

$$\cos \varphi = \frac{e^{j\varphi} + e^{-j\varphi}}{2}$$

$$\sin \varphi = \frac{e^{j\varphi} - e^{-j\varphi}}{2j}$$

it follows that

$$z = r(\cos \varphi + j \sin \varphi) = re^{j\varphi} \tag{17.1}$$

For r = 1, relation (17.1) is called ***Euler's formula***:

$$e^{j\varphi} = \cos \varphi + j \sin \varphi \tag{17.2}$$

Using formula (17.2), we find ***DeMoivre's formula***:

$$(\cos \varphi + j \sin \varphi)^n = (e^{j\varphi})^n = e^{jn\varphi} = \cos n\varphi + j \sin n\varphi \tag{17.3}$$

The polar form of complex numbers (17.1) is particularly useful to carry out the multiplication and division of complex numbers. For

$$z_1 = r_1\, e^{j\varphi_1}, \quad z_2 = r_2\, e^{j\varphi_2}$$

we obtain

$$z_1 z_2 = r_1 r_2 e^{j(\varphi_1 + \varphi_2)} = r_1 r_2 \{\cos(\varphi_1 + \varphi_2) + j \sin(\varphi_1 + \varphi_2)\}$$

and, similarly,

$$\frac{z_1}{z_2} = \frac{r_1}{r_2} e^{j(\varphi_1 - \varphi_2)} = \frac{r_1}{r_2}\{\cos(\varphi_1 - \varphi_2) + j \sin(\varphi_1 - \varphi_2)\}$$

Let n be a positive integer. We call the ***nth root*** of a complex number z, written $\sqrt[n]{z}$, any complex number w such that $w^n = z$. We disregard the case of $z = 0$. Setting

$$z = r\, e^{j\varphi}, \quad w = \sqrt[n]{z} = \rho\, e^{j\psi},$$

hence, the equality $w^n = z$ is written:

$$\rho^n e^{jn\psi} = r\, e^{j\varphi}$$

i.e., by equalizing the moduli and arguments,

$$\rho^n = r$$

$$n\,\psi = \varphi + 2\,k\,\pi$$

We, therefore, find

$$\rho = \sqrt[n]{r}$$

$$\psi = \frac{\varphi + 2\,k\,\pi}{n}, \quad k = 0, 1, 2, \ldots, n - 1$$

17.2 FUNCTIONS OF A COMPLEX VARIABLE

We now consider two complex variables z and w, and suppose that a relation is given so that, to each value in some region of the complex z-plane there is assigned a single value of w. Then, w is said to be a ***function*** of z, written

$$w = f(z)$$

Let u and v be the real and imaginary parts of w. Then, since w depends on $z = x + jy$, we have $w = f(z) = u(x,y) + jv(x,y)$, i.e., the real and imaginary parts of w are real functions of the real variables x and y.

The function $w = f(z)$ is called ***single-valued*** if for each value of z there is only onc value of w. If several values of w correspond to each value of z, the function f is ***multivalued***. The set of complex numbers which the function $w = f(z)$ can assume as z varies in the specified region of the z-plane, is called the ***range of values*** of the function $w = f(z)$ (fig. 17.2).

Let us proceed to the study of some elementary functions of a complex variable.

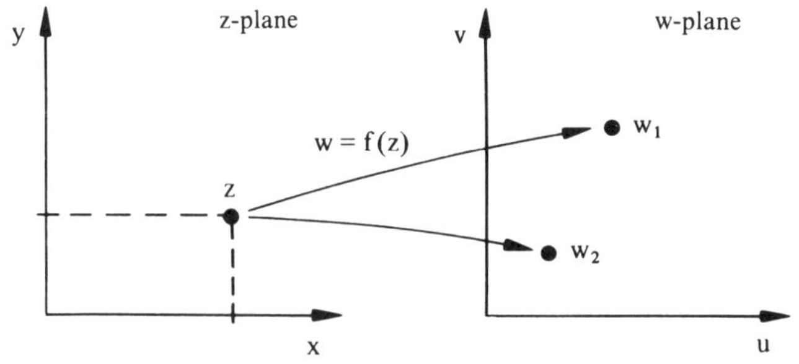

Fig. 17.2

17.2.1 Example: Quadratic function

Let $w = z^2$. Then,

$$w = u + jv = z^2 = (x + jy)^2 = x^2 - y^2 + 2jxy$$

The real part u of the quadratic function is, therefore, given by

$$u = x^2 - y^2$$

and the imaginary part v is given by

$$v = 2xy$$

For geometric interpretation, we shall determine the images of the straight lines $x = x_0$ and $y = y_0$ in the z-plane under the mapping defined by $w = z^2$. The image of the line $x = x_0$, $-\infty < y < \infty$, is given by the equations:

$$u = x_0^2 - y^2$$

$$v = 2\, x_0\, y$$

which is the parametric representation of the image in the w-plane.

Eliminating the parameter y, we find, for $x_0 \neq 0$:

$$u = x_0^2 - \frac{v^2}{4x_0^2}$$

i.e., the equation of a parabola with focus at the origin and opening to the left in the w-plane.

Similarly, the image of the line $y = y_0$, $-\infty < x < \infty$, is given the by the equations:

$$u = x^2 - y_0^2$$

$$v = 2xy_0$$

i.e., the equation of a parabola with focus at the origin and opening to the right in the w-plane. (fig. 17.3).

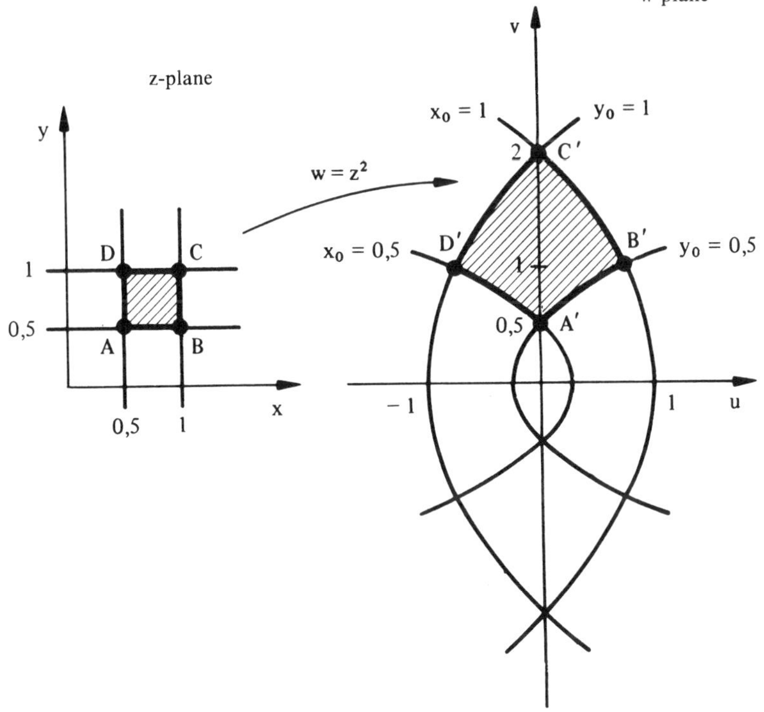

Fig. 17.3

17.2.2 Example: Inversion

Consider the function defined by

$$w = \frac{1}{z}$$

We have

$$w = u + jv = \frac{1}{z} = \frac{1}{x + jy} = \frac{x - jy}{x^2 + y^2}$$

thus,

$$u = \frac{x}{x^2 + y^2} \tag{17.4}$$

$$v = \frac{-y}{x^2 + y^2} \tag{17.5}$$

Equations (17.4) and (17.5) imply the relation:

$$u^2 + v^2 = \frac{x^2 + y^2}{(x^2 + y^2)^2} = \frac{1}{x^2 + y^2}$$

from which we obtain, again using (17.4) and (17.5):

$$x = \frac{u}{u^2 + v^2} \tag{17.6}$$

$$y = \frac{-v}{u^2 + v^2} \tag{17.7}$$

For geometric interpretation of the application $w = 1/z$, we first determine the image of an arbitrary straight line. Let us assume that the parametric representation is given in the following form, where λ designates the parameter:

$$\left.\begin{array}{l} x = x_0 + \lambda \cos\alpha \\ y = y_0 + \lambda \sin\alpha \end{array}\right\} -\infty < \lambda < +\infty$$

Substituting the variables x and y, according to (17.6) and (17.7), we obtain

$$\frac{u}{u^2 + v^2} = x_0 + \lambda \cos\alpha$$

$$\frac{-v}{u^2 + v^2} = y_0 + \lambda \sin\alpha$$

and eliminating the parameter λ:

$$\tan\alpha = \frac{\dfrac{-v}{u^2 + v^2} - y_0}{\dfrac{u}{u^2 + v^2} - x_0}$$

i.e.,

$$\sin\alpha \left(\frac{u}{u^2 + v^2} - x_0 \right) = \left(\frac{-v}{u^2 + v^2} - y_0 \right) \cos\alpha$$

Multiplying this preceding equation by u^2+v^2:

$$(u^2+v^2)(\cos\alpha\, y_0-\sin\alpha\, x_0)+u\sin\alpha+v\cos\alpha=0 \tag{17.8}$$

From equation (17.8), the image of a straight line for which $\cos\alpha\, y_0\neq\sin\alpha\, x_0$, i.e., which does not pass through the origin of the z-plane, therefore, is a circle passing through the origin of the w-plane. If $\cos\alpha\, y_0=\sin\alpha x_0$, i.e., if the straight line passes through the origin of the z-plane, its image will be a line passing through the origin of the w-plane.

Let us further determine the image of a circle of radius r, given in the form:

$$\left.\begin{aligned} x&=x_0+r\cos\varphi \\ y&=y_0+r\sin\varphi \end{aligned}\right\} 0\leq\varphi<2\pi$$

or, after elimination of the parameter φ:

$$(x-x_0)^2+(y-y_0)^2=r^2$$

Substituting the variables x and y, from (17.6) and (17.7), we find

$$r^2=\left(\frac{u}{u^2+v^2}-x_0\right)^2+\left(\frac{-v}{u^2+v^2}-y_0\right)^2$$

from which we get

$$\begin{aligned} r^2(u^2+v^2)^2&=[u-(u^2+v^2)x_0]^2+[-v-(u^2+v^2)y_0]^2 \\ &=u^2+v^2-2(ux_0-vy_0)(u^2+v^2)+(x_0^2+y_0^2)(u^2+v^2)^2 \end{aligned}$$

and dividing the preceding equation by u^2+v^2, assuming that $u^2+v^2\neq0$:

$$(r^2-x_0^2-y_0^2)(u^2+v^2)+2(ux_0-vy_0)=1 \tag{17.9}$$

From equations (17.9), the image of a circle for which $r^2-x_0^2-y_0^2\neq0$, i.e., which does not pass through the origin of the z-plane, therefore is, a circle in the w-plane. If $r^2-x_0^2-y_0^2=0$, i.e., if the circle passes through the origin of the z-plane, its image will be a straight line in the w-plane.

Thus, the transformation $w=1/z$ maps the totality of circles and straight lines in the z-plane onto circles and straight lines in the w-plane.

From the geometric point of view, the passage of the point z to the point $1/z$ is an ***inversion*** with respect to the unit circle followed by a ***symmetry*** with respect to the real axis. Figure (17.4) illustrates this passage for $z=0.5\, e^{j\pi/3}$; we have $w=1/z=2\,e^{-j\pi/3}$.

Let us here mention a practical example, taken from circuit theory, where the function $w=1/z$ is used. The complex impedance of a coil is given by

$$z=R+jL\omega \tag{17.10}$$

where R designates the resistance, L is the inductance, and ω is the angular frequency of the applied sinusoidal voltage. In the z-plane, all the points defined by equation (17.10) are located on the straight line parallel to the imaginary axis and passing at the distance R from the origin. If we are interested in the complex admittance $1/z$ of the coil, this straight line will be mapped onto the circle of radius $1/2R$ passing through the origin and centered on the x-axis (fig. 17.5).

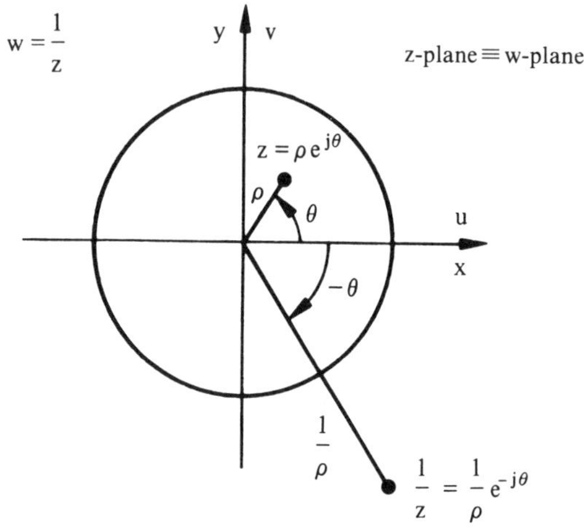

Fig. 17.4

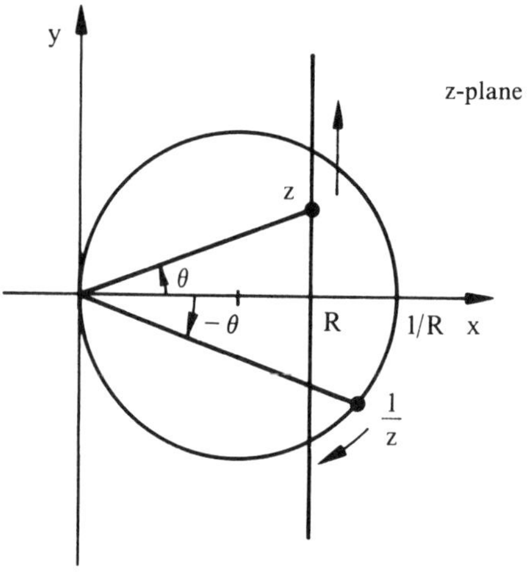

Fig. 17.5

17.2.3 Example: Bilinear transformation

The bilinear transformation is defined by

$$w = \frac{az + b}{cz + d}$$

where a, b, c, and d are arbitrary complex numbers such that $ad - bc \neq 0$.
This transformation is the combination of the following transformations:

$$w_1 = z + \frac{d}{c} \qquad \textit{:translation}$$

$$w_2 = \frac{1}{w_1} \qquad \textit{:inversion}$$

$$w_3 = \frac{bc - ad}{c^2} w_2 \qquad \textit{:stretching and rotation}$$

$$w = w_3 + \frac{a}{c} \qquad \textit{:translation}$$

namely,

$$w = \frac{bc - ad}{c^2} w_2 + \frac{a}{c} = \frac{bc - ad}{c^2} \frac{1}{w_1} + \frac{a}{c} = \frac{bc - ad}{c^2} \frac{1}{z + \frac{d}{c}} + \frac{a}{c}$$

$$= \frac{bc - ad + a(cz + d)}{c(cz + d)} = \frac{az + b}{cz + d}$$

Because each translation, inversion, stretching, and rotation maps the totality of circles and straight lines onto circles and straight lines, the bilinear function has the same property.

17.2.4 Example: Exponential function

The exponential function is defined by

$$w = e^z$$

We have

$$w = e^z = e^{x+jy} = e^x (\cos y + j \sin y)$$

For geometric interpretation, we are interested in the images of the straight lines $x = x_0$ and $y = y_0$.
If $x = x_0$, then

$$w = u + jv = e^{x_0}(\cos y + j \sin y)$$

from which we get

$$u = e^{x_0}\cos y$$
$$v = e^{x_0}\sin y,$$

or eliminating the parameter y, we have

$$u^2 + v^2 = e^{2x_0}$$

i.e., the equation for a circle of radius e^{x_0} centered at the origin.

For $y = y_0$, we obtain

$$w = u + jv = e^{x}(\cos y_0 + j \sin y_0)$$

from which we get

$$u = e^{x}\cos y_0$$
$$v = e^{x}\sin y_0$$

or, eliminating the parameter x, we have

$$\frac{v}{u} = \tan y_0$$

which represents a straight line that passes through the origin of the w-plane and forms the angle y_0 with the u-axis (fig. 17.6).

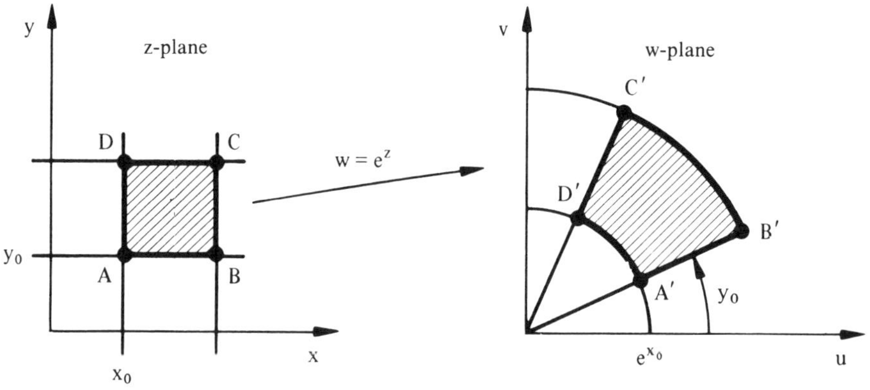

Fig. 17.6

For any integer k, we have

$$e^{z + j2k\pi} = e^{z}\, e^{j2k\pi} = e^{z}$$

i.e., the exponential function is ***periodic*** with period $2\pi j$.

17.2.5 Example: Natural logarithm

The natural logarithm is defined as the inverse of the exponential function:

$$w = \ln z \Longleftrightarrow z = e^{w}$$

Let r and φ be the polar coordinates of z, $z = r\,e^{j\varphi}$ and $w = u + jv$. Then,

$$z = r\,e^{j\varphi} = e^{w} = e^{u+jv} = e^{u}e^{jv}$$

from which we get

$r = e^{u}$, which implies $u = \ln r$

and

$e^{j\varphi} = e^{jv}$, which implies $v = \varphi + 2k\pi$, k integer. Thus, we have

$$\ln z = w = u + jv = \ln r + j\,(\varphi + 2\,k\,\pi),$$

from which we get

$$\ln z = \ln|z| + j\,(\arg z + 2k\pi), \text{ k integer}$$

The logarithm is ***multivalued***; we obtain its ***principal value*** for $k = 0$ (fig. 17.7), which is the real natural logarithm from elementary calculus if z is real and positive.

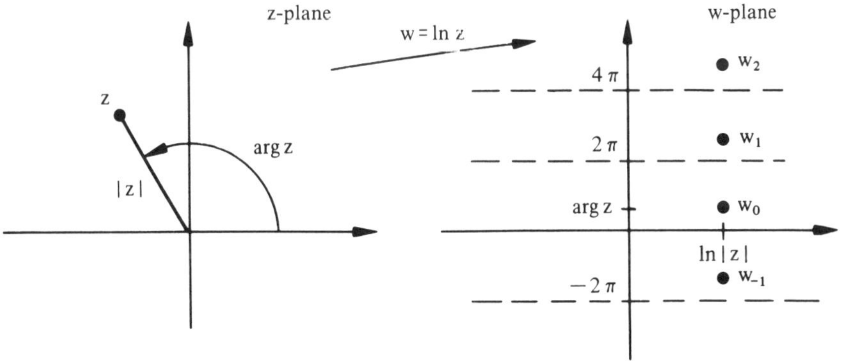

Fig. 17.7

Chapter 18

Analytic Functions

18.1 DEFINITION OF THE DERIVATIVE

Let f(z) be a function of a complex variable defined in some region D of the z-plane:

$$f(z) = u + jv$$

The function f(z) is said to be ***continuous*** at the point $z_0 \in D$, if $f(z_0 + \Delta z)$ approaches $f(z_0)$ as Δz approaches zero from any direction. We can easily see that f(z) is continuous in $z_0 = x_0 + jy_0$ if and only if the real and imaginary parts of f(z), i.e.,

$$u = u(x, y) \text{ and } v = v(x, y)$$

are continuous at the point (x_0, y_0).

We call f(z) ***differentiable*** at the point $z_0 \in D$, if the quotient:

$$\frac{f(z_0 + \Delta z) - f(z_0)}{\Delta z}$$

has one and only one limit as Δz approaches zero from any direction. This limit is said to be the ***derivative*** of f(z) at the point z_0 and is designated by $df(z_0)/dz$, or $f'(z_0)$. If f(z) is differentiable at each point of the region D, then f(z) is said to be ***analytic*** in D.

18.2 CAUCHY-RIEMANN CONDITIONS

Let us assume that the real and imaginary parts of the function:

$$f(z) = u(x, y) + jv(x, y)$$

are continuously differentiable with respect to x and y in a domain D of the z-plane.

THEOREM: The function $f(z)=u(x, y)+jv(x, y)$ is analytic in D if and only if the following conditions are satisfied:

$$\frac{\partial u}{\partial x}=\frac{\partial v}{\partial y}, \qquad \frac{\partial u}{\partial y}=-\frac{\partial v}{\partial x} \tag{18.1}$$

These conditions are called the ***Cauchy-Riemann conditions***.

To demonstrate this theorem, we first verify that the conditions (18.1) are *necessary*. We, therefore, assume the existence of the limit:

$$f'(z)=\frac{df}{dz}(z)=\lim_{\Delta z\to 0}\frac{f(z+\Delta z)-f(z)}{\Delta z}$$

where Δz approaches zero from any direction. If Δz approaches zero along the real axis, i.e., $\Delta z=\Delta x$, we find

$$\frac{df}{dz}=\frac{df}{dx}\bigg|_{dy=0}=\frac{\partial f}{\partial x}=\frac{\partial u}{\partial x}+j\frac{\partial v}{\partial x} \tag{18.2}$$

and if approaching zero along the imaginary axis, i.e., $\Delta z=j\Delta y$, we obtain

$$\frac{df}{dz}=\frac{1}{j}\frac{df}{dy}\bigg|_{dx=0}=\frac{1}{j}\frac{\partial f}{\partial y}=\frac{1}{j}\left(\frac{\partial u}{\partial y}+j\frac{\partial v}{\partial y}\right) \tag{18.3}$$

Comparing relations (18.2) and (18.3), we have

$$\frac{\partial u}{\partial x}+j\frac{\partial v}{\partial x}=\frac{\partial v}{\partial y}-j\frac{\partial u}{\partial y}$$

from which we get the equations (18.1).

Next, we demonstrate that the Cauchy-Riemann conditions (18.1) are also sufficient. The exact differential df of the function $f(z)$ has the form:

$$df=\left(\frac{\partial u}{\partial x}+j\frac{\partial v}{\partial x}\right)dx+\left(\frac{\partial u}{\partial y}+j\frac{\partial v}{\partial y}\right)dy$$

Substituting $\partial u/\partial y$ and $\partial v/\partial y$ into the second term on the right side of the preceding equation, according to equations (18.1), we find

$$df=\left(\frac{\partial u}{\partial x}+j\frac{\partial v}{\partial x}\right)dx+\left(-\frac{\partial v}{\partial x}+j\frac{\partial u}{\partial x}\right)dy$$

$$=\left(\frac{\partial u}{\partial x}+j\frac{\partial v}{\partial x}\right)(dx+j\,dy)=\frac{\partial f}{\partial x}dz$$

Thus, the limit df/dz exists and takes the value $\partial f/\partial x$, a value that is independent of the direction from which Δz approaches zero. Therefore, the theorem is demonstrated.

It is easily seen that the well known laws of differentiation are maintained: if f(z) and g(z) are two analytic functions of a complex variable, then

$$(f(z)+g(z))' = f'(z)+g'(z)$$
$$(f(z)\,g(z))' = f'(z)\,g(z)+f(z)\,g'(z)$$
$$\left(\frac{f(z)}{g(z)}\right)' = \frac{f'(z)\,g(z)-f(z)\,g'(z)}{g(z)^2} \quad \text{for } g(z)\neq 0$$

If $f(z)=u(x, y)+jv(x, y)$ is an analytic function, we have seen that the functions u and v satisfy the Cauchy-Riemann conditions (18.1), which immediately imply that u and v are solutions of ***Laplace's equation***:

$$\frac{\partial^2 u}{\partial x^2}+\frac{\partial^2 u}{\partial y^2}=0, \quad \frac{\partial^2 v}{\partial x^2}+\frac{\partial^2 v}{\partial y^2}=0 \tag{18.4}$$

A solution of Laplace's equation is called a *harmonic function*. Hence, the real and imaginary parts of an analytic function are said to be ***conjugate harmonic*** functions.

By the same token, if u is a harmonic function, then the conjugate harmonic function v can be determined from the Cauchy-Riemann conditions as follows. The differential form:

$$\frac{\partial v}{\partial x}dx+\frac{\partial v}{\partial y}dy=-\frac{\partial u}{\partial y}dx+\frac{\partial u}{\partial x}dy$$

is the exact differential dv of a function v, since the right side of this equation satisfies the condition:

$$\frac{\partial}{\partial y}\left(-\frac{\partial u}{\partial y}\right)=\frac{\partial}{\partial x}\left(\frac{\partial u}{\partial x}\right)$$

which follows from the fact that u is harmonic. If the harmonic function u is defined in a simply connected region, then, according to Green's theorem in the plane, the line integral:

$$v(x,y)=\int_{(x_0,y_0)}^{(x,y)}\left(-\frac{\partial u}{\partial y}dx+\frac{\partial u}{\partial x}dy\right) \tag{18.5}$$

is independent of the path joining the fixed initial point (x_0, y_0) and the varying end point (x, y). The line integral satisfies the Cauchy-Riemann conditions (18.1):

$$\frac{\partial v}{\partial x}=-\frac{\partial u}{\partial y}, \quad \frac{\partial v}{\partial y}=\frac{\partial u}{\partial x}$$

and we have

$$\frac{\partial^2 v}{\partial x^2}+\frac{\partial^2 v}{\partial y^2}=-\frac{\partial}{\partial x}\left(\frac{\partial u}{\partial y}\right)+\frac{\partial}{\partial y}\left(\frac{\partial u}{\partial x}\right)=-\frac{\partial^2 u}{\partial x\partial y}+\frac{\partial^2 u}{\partial x\partial y}=0$$

Thus, the function v, as defined by equation (18.5), is the conjugate harmonic function of u: it is single-valued in a simply connected region and determined to within an additive constant.

As an example, we shall consider the exponential function:

$$f(z)=e^z=e^{x+jy}=e^x(\cos y+\sin y)$$

The real and imaginary parts:

$$u=e^x \cos y$$
$$v=e^x \sin y$$

satisfy the Cauchy-Riemann conditions (18.1) for all z:

$$\frac{\partial u}{\partial x}=e^x \cos y=\frac{\partial v}{\partial y}$$

$$\frac{\partial u}{\partial y}=-e^x \sin y=-\frac{\partial v}{\partial x}$$

The exponential function is, therefore, analytic for all z, and its derivative is equal to itself:

$$f'(z)=\frac{\partial f}{\partial x}=\frac{\partial}{\partial x}(e^x(\cos y+j\sin y))=e^z$$

It is easily verified that u an v are harmonic functions:

$$\Delta u=\frac{\partial^2 u}{\partial x^2}+\frac{\partial^2 u}{\partial y^2}=e^x \cos y-e^x \cos y=0$$

$$\Delta v=\frac{\partial^2 v}{\partial x^2}+\frac{\partial^2 v}{\partial y^2}=e^x \sin y-e^x \sin y=0$$

Conversely, if we wish to determine an analytic function f(z) such that its real part is the given harmonic function:

$$u=e^x \cos y$$

we proceed as follows. The differential form:

$$-\frac{\partial u}{\partial y}dx+\frac{\partial u}{\partial x}dy=e^x \sin y\, dx+e^x \cos y\, dy$$

is the exact differential dv of the conjugate harmonic function v of u. We have (18.5):

$$v(x,y) = \int_{(0,0)}^{(x,y)} (e^x \sin y\, dx + e^x \cos y\, dy) + C$$

We evaluate this line integral by integrating, first, along the y-axis from (0,0) to (0,y), and then from (0,y) to (x,y) along the line parallel to the x-axis:

$$v(x,y) = \int_0^y \cos y\, dy + \sin y \int_0^x e^x dx + C$$

$$= \sin y + \sin y\, e^x - \sin y + C$$
$$= e^x \sin y + C$$

Hence, the most general analytic function with the specified real part is the exponential function to within an additive constant:

$$f(z) = e^x \cos y + j\, e^x \sin y + C = e^{x+jy} + C = e^z + C$$

18.3 CONFORMAL MAPPING

We shall now consider the most important geometrical property of the mappings defined by analytic functions, namely, their conformality.

A mapping in the plane is ***conformal*** if it preserves angles in magnitude as well as in sense.

The following theorem is fundamental.

THEOREM: The mapping defined by an analytic function $w = f(z)$ is conformal, except at points where the derivative $f'(z)$ is zero.

To demonstrate this theorem, we consider two curves C_1 and C_2 in the z-plane, which intersect at the point z_0. We designate their images in the w-plane by C_1^* and C_2^*. The curves C_1^* and C_2^* intersect at the image point $w_0 = f(z_0)$ (fig. 18.3). We shall demonstrate that the images of these two intersecting curves make the same angle of intersection as the two curves in the z-plane, both in magnitude and in direction. With this aim, we first consider the curve C_1, which is represented in the form:

$$C_1: z = z(t)$$

where t is a real parameter, Thus, we have for its image C_1^* the following representation:

$$C_1^*: w = w(t) = f(z(t))$$

The vector tangent to C_1 at $z_0 = z(t_0)$ is given by

$$\frac{dz}{dt}(t_0)$$

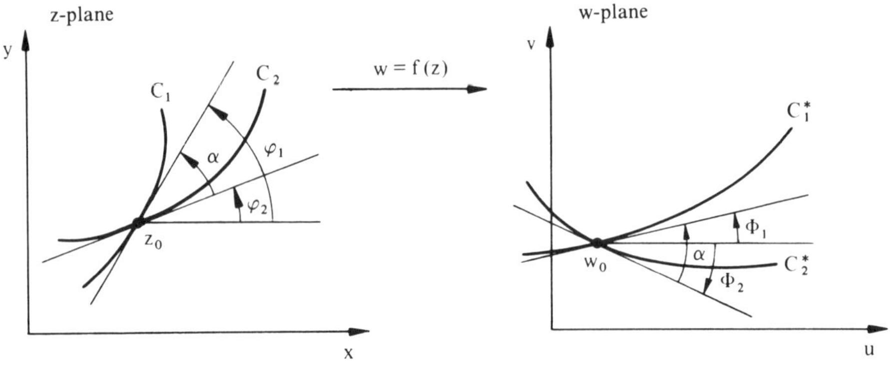

Fig. 18.1

and the vector tangent to C_1^* at $w_0 = f(z(t_0))$ by

$$\frac{dw}{dt}(t_0) = \frac{df}{dz}\frac{dz}{dt}(t_0) = f'(z_0)\frac{dz}{dt}(t_0) \tag{18.6}$$

If we set (fig. 18.1)

$$\frac{dz}{dt}(t_0) = r_0\, e^{j\varphi_1}$$

$$\frac{dw}{dt}(t_0) = \rho_0\, e^{j\Phi_1}$$

and for the derivative f′ at the point z_0:

$$f'(z_0) = R\, e^{j\psi}$$

then, relation (18.6) is written in the following manner:

$$\rho_0 e^{j\Phi_1} = R\, r_0\, e^{j(\varphi_1 + \psi)} \tag{18.7}$$

Since we assume that $f'(z_0) \neq 0$, we have $R \neq 0$, and according to equation (18.16), we find

$$\Phi_1 = \varphi_1 + \psi\, (\text{mod } 2\pi)$$

where the angle ψ depends only on the derivative f′ at the point z_0.

We now consider the second curve C_2. Similarly, we find

$$\Phi_2 = \varphi_2 + \psi\, (\text{mod } 2\pi)$$

Since $\alpha = \varphi_2 - \varphi_1 = \Phi_2 - \Phi_1$, the angle between the tangents in the z-plane and their images in the w-plane is preserved in magnitude as well as in sense under the mapping $w = f(z)$ (fig. 18.1). The theorem is thus demonstrated.

Conformal mappings play an important role in the solution of many practical problems, especially in the field of electrostatics, hydrodynamics, aerodynamics, and elasticity. The method consists of using a conformal mapping that maps a given region into another which is more suitable for the solution of the problem. This powerful method is also justified by the fact that a harmonic function remains harmonic under a change of the variables arising from a conformal mapping $w = f(z)$.

We shall proceed to study two conformal transformations which are of great practical interest.

18.3.1 Example: Joukowski's transformation

Joukowski's transformation is defined by

$$w = f(z) = \frac{1}{2}\left(z + \frac{1}{z}\right) \tag{18.8}$$

It is defined for all the points $z \neq 0$, and since its derivative equals

$$f'(z) = \frac{1}{2}\left(1 - \frac{1}{z^2}\right)$$

the mapping is conformal except at the points $z = 1$ and $z = -1$, which are fixed points of the transformation.

Representing the unit circle in parametric form:

$$z = e^{j\varphi}, \quad 0 \leq \varphi < 2\pi$$

we obtain the following expression for its image in the w-plane:

$$w = \frac{1}{2}(e^{j\varphi} + e^{-j\varphi}) = \cos\varphi, \quad 0 \leq \varphi < 2\pi$$

If the point z traverses the unit circle, its image then traverses the line segment $[-1, +1]$.

The circles $|z| = R \neq 1$, $R > 0$, are mapped onto ellipses:

$$w = u + jv = \frac{1}{2}\left(Re^{j\varphi} + \frac{1}{Re^{j\varphi}}\right)$$

$$= \frac{1}{2}\left(R + \frac{1}{R}\right)\cos\varphi + \frac{j}{2}\left(R - \frac{1}{R}\right)\sin\varphi$$

from which we get

$$u = \frac{1}{2}\left(R + \frac{1}{R}\right)\cos\varphi \qquad (18.9)$$

$$v = \frac{1}{2}\left(R - \frac{1}{R}\right)\sin\varphi \qquad (18.10)$$

and eliminating φ:

$$\frac{u^2}{a^2} + \frac{v^2}{b^2} = 1 \qquad (18.11)$$

with

$$a = \frac{1}{2}\left(R + \frac{1}{R}\right) \text{ and } b = \frac{1}{2}\left(R - \frac{1}{R}\right)$$

From these expressions for the principle half-axes, we see that the two circles with radii R and $1/R$ map onto the same ellipse. The lines φ_0 = constant are mapped onto hyperbolas. In effect, according to relations (18.9) and (18.10), we have

$$\frac{4u^2}{\cos^2\varphi_0} = R^2 + 2 + \frac{1}{R^2}$$

$$\frac{4v^2}{\sin^2\varphi_0} = R^2 - 2 + \frac{1}{R^2}$$

and taking the difference of the two preceding equalities, we find

$$\frac{u^2}{\cos^2\varphi_0} - \frac{v^2}{\sin^2\varphi_0} = 1 \qquad (18.12)$$

Since the mapping is conformal except at the points $z = 1$ and $z = -1$, the hyperbolas are the orthogonal trajectories of the ellipses. (fig. 18.2).

Joukowski's transformation (18.8) tranforms circles passing through $z = 1$ and encircling $z = -1$ into curves with sharp trailing edges that resemble ***airfoils*** (fig. 18.3), known as Joukowski airfoils.

18.3.2 Example: Bilinear transformation

In this example, we shall consider the following bilinear transformation:

$$w = f(z) = \frac{1 + z}{1 - z} \qquad (18.13)$$

Since its derivative is

$$f'(z) = \frac{2}{(1 - z)^2}$$

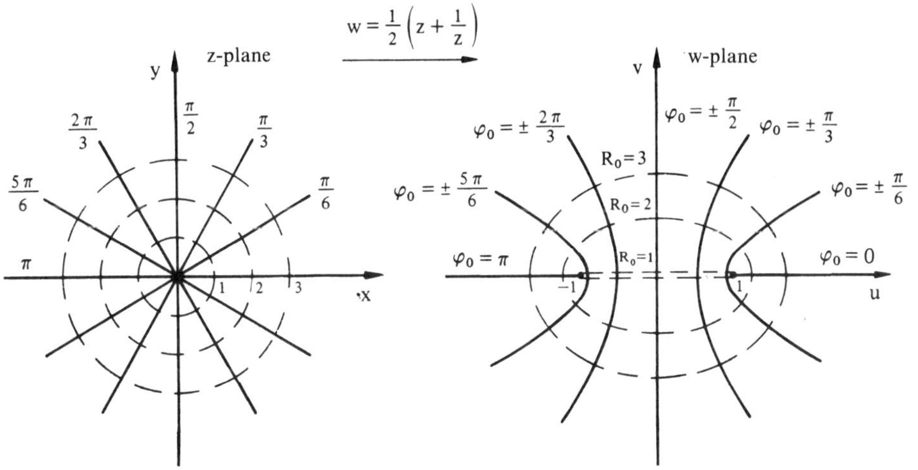

Fig. 18.2

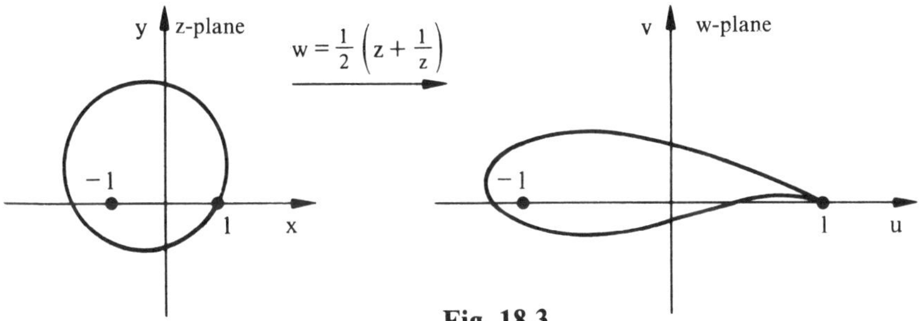

Fig. 18.3

the mapping is conformal except at the point $z = 1$.

The point $z = 1$ is a pole of the transformation, i.e., its image under the mapping is the point at infinity.

We are interested in the images of circles centered on the x-axis and passing through the point $z = 1$. We mentioned in section 17.2.3 that each bilinear transformation maps circles and straight lines into circles and straight lines. Given that the point $z = 1$ maps onto the point at infinity, each circle passing through $z = 1$ must be mapped into a straight line. The x-axis passes through the points $z = 0$, $z = 1$, and $z = -1$, which are respectively transformed into $w = 1$, $w = \infty$, and $w = 0$. The image of the x-axis, therefore, is the u-axis. The circles centered on the x-axis intersect the x-axis at right angles, and since the transformation (18.13) is conformal for all the $z \neq 1$, the images of these circles must also intersect the u-axis at right angles. The images of the circles centered on the x-axis passing through z = 1, therefore, are parallel lines, perpendicular to the u-axis (fig. 18.4).

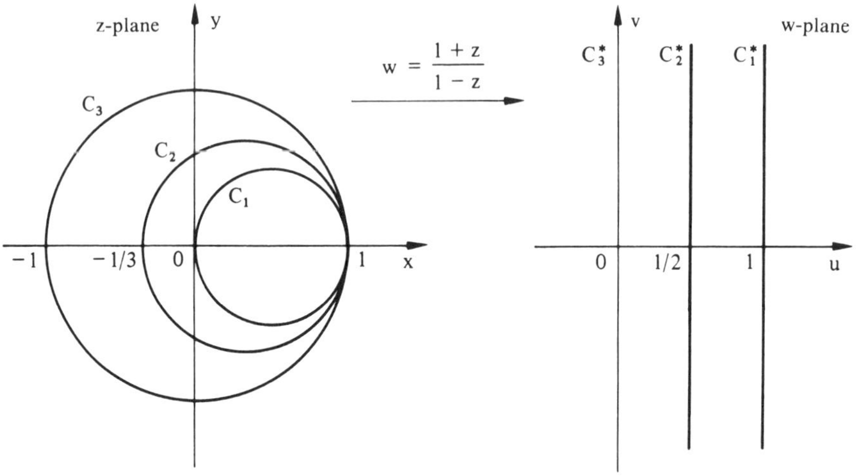

Fig. 18.4

18.4 CAUCHY-RIEMANN CONDITIONS IN POLAR COORDINATES

Let us assume that the real and imaginary parts of the function w = f(z) are expressed in terms of polar coordinates.

$$f(z) = u(r,\varphi) + jv(r,\varphi) \quad \text{where } z = r\,e^{j\varphi}$$

If f(z) is analytic, the quotient:

$$\frac{f(z+\Delta z) - f(z)}{\Delta z}$$

has one and only one limit as Δz approaches zero from any direction. Since $dz = e^{j\varphi}(dr + jr\,d\varphi)$, we find, if Δz approaches zero along the position vector of z, i.e., $d\varphi = 0$:

$$\frac{df}{dz} = e^{-j\varphi} \left.\frac{df}{dr}\right|_{d\varphi=0} = e^{-j\varphi}\left(\frac{\partial u}{\partial r} + j\frac{\partial v}{\partial r}\right) \tag{18.14}$$

Also, if Δz approaches zero along the tangent to the circle of radius r centered at the origin, i.e., $dr = 0$, we obtain

$$\frac{df}{dz} = \frac{e^{-j\varphi}}{jr} \left.\frac{df}{d\varphi}\right|_{dr=0} = \frac{e^{-j\varphi}}{jr}\left(\frac{\partial u}{\partial \varphi} + j\frac{\partial v}{\partial \varphi}\right) \tag{18.15}$$

Equating the expressions (18.14) and (18.15) for the derivative df/dz, we arrive at the ***Cauchy-Riemann conditions in polar form***:

$$\frac{\partial u}{\partial r} = \frac{1}{r}\frac{\partial v}{\partial \varphi}$$
$$\frac{\partial v}{\partial r} = -\frac{1}{r}\frac{\partial u}{\partial \varphi} \qquad (18.16)$$

According to relation (18.14), the derivative equals

$$\frac{df}{dz} = e^{-j\varphi}\left(\frac{\partial u}{\partial r} + j\frac{\partial v}{\partial r}\right) = \frac{r}{z}\frac{\partial f}{\partial r}$$

As an example, we consider the principal value of the natural logarithm:

$$f(z) = \ln z = \ln r\, e^{j\varphi} = \ln r + j\varphi$$

The real and imaginary parts:

$$u = \ln r$$
$$v = \varphi$$

obviously satisfy the Cauchy-Riemann conditions (18.16) at any point $z \neq 0$, and their derivatives equal

$$\frac{\partial u}{\partial r} = \frac{1}{r} = \frac{1}{r}\frac{\partial v}{\partial \varphi}$$
$$\frac{\partial v}{\partial r} = 0 = -\frac{1}{r}\frac{\partial u}{\partial \varphi}$$

Therefore, the logarithmic function is analytic, except at the point $z = 0$, and its derivative equals

$$f'(z) = \frac{r}{z}\frac{1}{r} = \frac{1}{z}$$

Chapter 19

Complex Integrals

19.1 LINE INTEGRAL IN THE COMPLEX PLANE

Let $f(z) = u + jv$ be a continuous function of a complex variable, which is defined at each point of a smooth curve C in the complex plane. Then, the ***line integral*** of $f(z)$ along the curve C is defined by the limit

$$\int_C f(z)\,dz = \lim_{n\to\infty} \sum_{i=0}^{n-1} f(\xi_i)(z_{i+i} - z_i) \tag{19.1}$$

where the points $z_0 = a, z_1, \ldots, z_n = b$ divide C into n arcs between the end points a and b, ξ_i is an arbitrary point of the chord $[z_i, z_{i+1}]$ of C, and the limit is taken such that the greatest chord length

$$\max_{i=0,\ldots,n-1} |z_{i+1} - z_i|$$

approaches zero as n approaches ∞ (fig. 19.1).

Integral (19.1) can be expressed in terms of two real line integrals in the following way:

$$\int_C f(z)\,dz = \int_C (u + jv)(dx + j\,dy)$$

$$= \int_C (u\,dx - v\,dy) + j \int_C (v\,dx + u\,dy) \tag{19.2}$$

If C is a ***closed curve***, the line integral along C is designated by

$$\oint_C f(z)\,dz$$

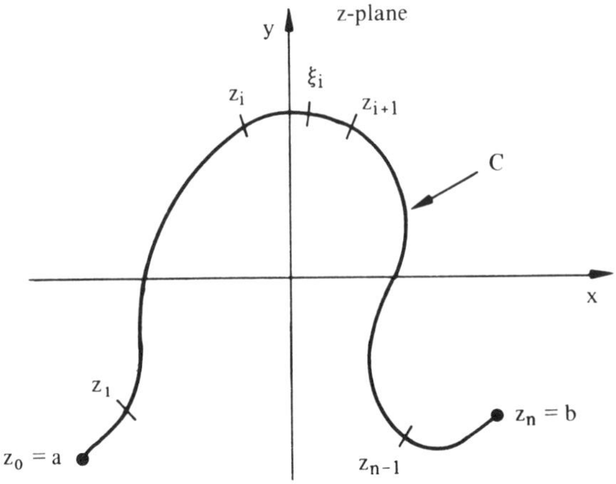

Fig. 19.1

19.1.1 Example: Line integral of a non-analytic function

We wish to evaluate the line integral:

$$\int_C \bar{z}\, dz$$

along the curve C:

$$z(t) = t^2 + jt, \quad 0 \leq t \leq 2$$

which joins the points $a = z(0) = 0$ and $b = z(2) = 4 + j2$ (fig. 19.2).

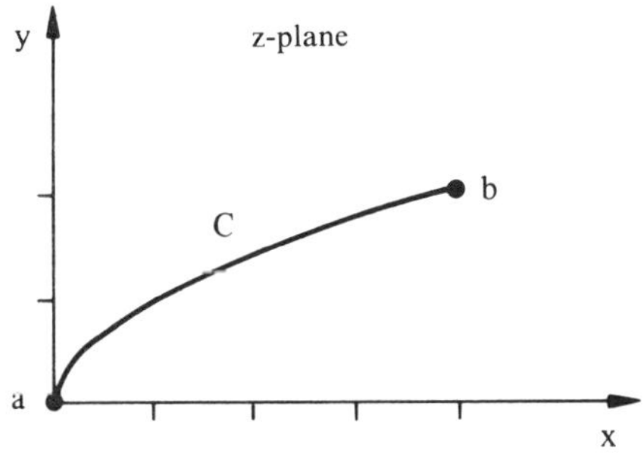

Fig. 19.2

According to formula (19.2), we calculate

$$\int_C \bar{z}\,dz = \int_0^2 (t^2 - jt)(2t + j)\,dt = \int_0^2 (2t^3 + t)\,dt - j\int_0^2 t^2\,dt = 10 - j\frac{8}{3}$$

19.1.2 Example: Line integral of the *n*th power on the unit circle

We wish to evaluate the line integral:

$$\oint_C z^n dz, \quad n \text{ integer}$$

where the path C:

$$z(\varphi) = e^{j\varphi}, \quad 0 \leq \varphi \leq 2\pi$$

is the unit circle.

For $n = -1$, we have

$$\oint_C^{\infty} \frac{1}{z} dz = \int_0^{2\pi} \frac{je^{j\varphi}}{e^{j\varphi}}\,d\varphi = j\int_0^{2\pi} d\varphi = 2\pi j \tag{19.3}$$

For $n \neq -1$, we calculate

$$\oint_C z^n\,dz = \int_0^{2\pi} e^{jn\varphi} je^{j\varphi}\,d\varphi = j\int_0^{2\pi} e^{(n+1)j\varphi}\,d\varphi$$

$$= j\,\frac{e^{(n+1)j\varphi}}{(n+1)j}\Bigg|_0^{2\pi} = \frac{1}{n+1}\,[e^{(n+1)j2\pi} - 1] = 0 \tag{19.4}$$

19.2 CAUCHY'S INTEGRAL THEOREM

Cauchy's integral theorem is very important in complex analysis and has various theoretical and practical consequences.

Let f(z) be an analytic function in a simply connected domain D and let C be a simple closed path in D (fig. 19.3). Then, ***Cauchy's theorem*** states:

$$\oint_C f(z)\,dz = 0 \tag{19.5}$$

To demonstrate this theorem, we represent the complex line integral in terms of two real integrals (19.2):

$$\oint_C f(z)\,dz = \oint_C (u+jv)(dx+j\,dy)$$
$$= \oint_C (u\,dx - v\,dy) + j \oint_C (v\,dx + u\,dy) \tag{19.6}$$

Applying Green's formula (11.2) to each integral on the right side of equation (19.6), we find

$$\oint_C (u\,dx + (-v)\,dy) = \iint_{D_1} \left\{ \frac{\partial}{\partial x}(-v) - \frac{\partial u}{\partial y} \right\} dx\,dy$$
$$= -\iint_{D_1} \left(\frac{\partial v}{\partial x} + \frac{\partial u}{\partial y} \right) dx\,dy = 0 \tag{19.7}$$

$$\oint_C (v\,dx + u\,dy) = \iint_{D_1} \left(\frac{\partial u}{\partial x} - \frac{\partial v}{\partial y} \right) dx\,dy = 0 \tag{19.8}$$

Since f(z) is analytic, its real and imaginary part satisfy the Cauchy-Riemann conditions (18.1), and so the above integrals are zero. This completes the proof.

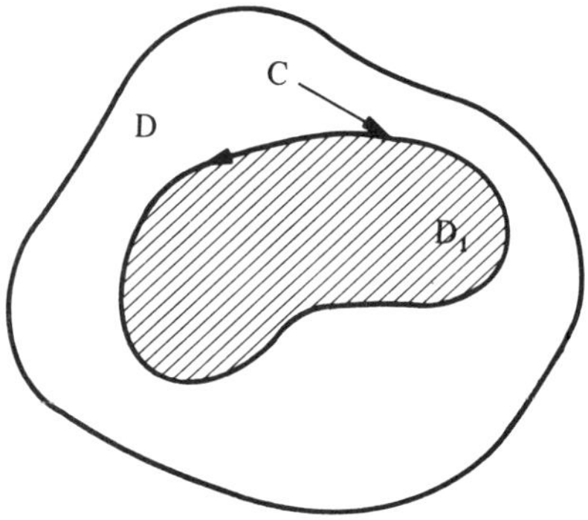

Fig. 19.3

Cauchy's theorem has an important ***corollary***, as given below.

Let f(z) be an analytic function in a simply connected region D. If a and b are any two points of D, then the line integral:

$$\int_a^b f(z)\,dz$$

is independent of the path in D that joins the points a and b (fig. 19.4)

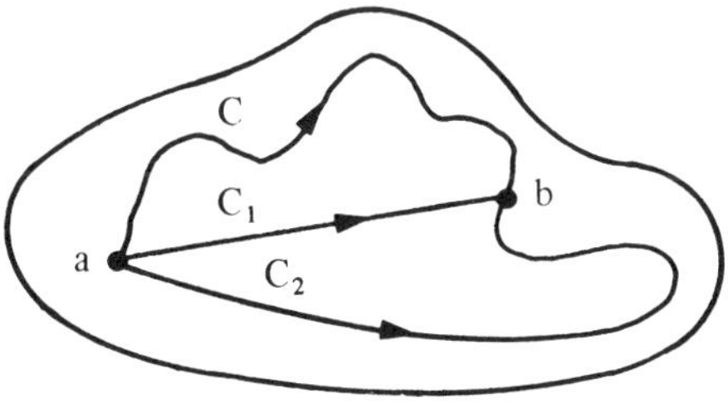

Fig. 19.4

The demonstration is as follows. If we consider two distinct paths C_1 and C_2 joining a and b, then the integral on the closed path formed by those two paths is zero. Hence, the two line integrals are equal and independent of the path joining the two points.

19.2.1 Example: Verfication of Cauchy's theorem

We shall consider Joukowski's function:

$$f(z) = \frac{1}{2}\left(z + \frac{1}{z}\right)$$

and we wish to determine its line integral along three different paths C_i, i = 1, 2, 3, joining the points a = 1 and b = j (fig. 19.5).•

For the integral along C_1, we calculate

$$\frac{1}{2}\int_{C_1}\left(z+\frac{1}{z}\right)dz = \frac{1}{2}\left[\int_0^1\left(1+jy+\frac{1}{1+jy}\right)j\,dy + \int_1^0\left(x+j+\frac{1}{x+j}\right)dx\right]$$

$$\frac{1}{2}\int_{C_1}\left(z+\frac{1}{z}\right)dz = \frac{1}{2}\left(-1+j\frac{\pi}{2}\right) = -\frac{1}{2}+j\frac{\pi}{4}$$

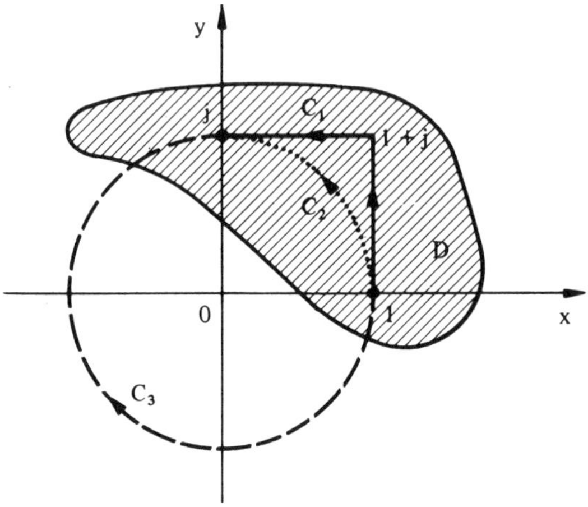

Fig. 19.5

Joukowski's function is analytic in the region bounded by C_1 and C_2 (fig. 19.5). Thus, according to Cauchy's theorem the integral along C_2 takes the same value as that along C_1 i.e.,

$$\frac{1}{2}\int_{C_2}\left(z+\frac{1}{z}\right)dz = -\frac{1}{2}+j\frac{\pi}{4}$$

To calculate the integral along C_3, we choose the polar representation of the curve C_3:

$$\frac{1}{2}\int_{C_3}\left(z+\frac{1}{z}\right)dz = \frac{1}{2}\int_0^{-3\pi/2}(e^{j\varphi}+e^{-j\varphi})\,j e^{j\varphi}\,d\varphi$$

$$= \frac{1}{2}\left(\frac{e^{2j\varphi}}{2}+j\varphi\right)\Bigg|_0^{-3\pi/2} = \frac{1}{2}\left(-1-j\frac{3\pi}{2}\right) = -\frac{1}{2}-j\frac{3\pi}{4}$$

Thus, the integral along C_3 differs from the integral along C_1, although the two paths join the same points $a = 1$ and $b = j$. This is due to the fact that Joukowski's function is not analytic at $z = 0$. Hence, Cauchy's theorem is not valid for the unit circle.

19.3 INDEFINITE INTEGRAL OF AN ANALYTIC FUNCTION

Let $f(z)$ be an analytic function in a simply connected domain D. Then, the function $F(z)$ is said to be an indefinite integral or primitive of $f(z)$ if

$$F'(z) = f(z)$$

at every point z of D.

The indefinite integral of $f(z)$ can be represented in the form

$$F(z) = \int_a^z f(z)\,dz \tag{19.9}$$

where the line integral is taken along any path located entirely in D and joining the fixed point a and z. According to Cauchy's theorem, the integral (19.9) is independent of the path and depends only on the end point z.

In order to show that the indefinite integral thus defined is analytic in D and that its derivative is equal to $f(z)$, we first set

$$F(z) = U(x,y) + jV(x,y)$$

Approaching the point z along a line parallel to the real axis, i.e., $\Delta z = \Delta x$, we calculate for the derivative:

$$\left.\frac{dF}{dx}\right|_{dy=0} = \frac{\partial U}{\partial x} + j\frac{\partial V}{\partial x} = \lim_{\Delta x \to 0}\left[\frac{1}{\Delta x}\int_{\substack{x \\ y=\text{const.}}}^{x+\Delta x} f(z)\,dx\right] = f(z) \tag{19.10}$$

and approaching z along a line parallel to the imaginary axis, i.e., $\Delta z = j\Delta y$, we obtain

$$\left.\frac{1}{j}\frac{dF}{dy}\right|_{dx=0} = -j\left(\frac{\partial U}{\partial y} + j\frac{\partial V}{\partial y}\right)$$

$$= \lim_{\Delta y \to 0}\left[\frac{1}{j\Delta y}\int_{\substack{y \\ x=\text{const.}}}^{y+\Delta x} f(z)\,j\,dy\right] = f(z) \tag{19.11}$$

Relations (19.10) and (19.11) immediately imply the validity of the Cauchy-Riemann conditions (18.1) for U and V. Thus, F(z) is an analytic function in D (section 18.2), and we have

$$F'(z)=f(z)$$

at any point z of D.

It remains to be seen that the indefinite integral of an analytic function f(z) is determined to within an additive constant. With this aim, let $F_1(z)$ and $F_2(z)$ be two primitives of f(z), that is,

$$F_1(z)=\int_a^z f(z)\,dz, \quad F_2(z)=\int_b^z f(z)\,dz$$

where $a \neq b$. Thus, the difference:

$$G(z)=U(x,y)+jV(x,y)=F_1(z)-F_2(z)$$

is an analytic function, and

$$G'(z)=\frac{\partial U}{\partial x}+j\frac{\partial V}{\partial x}=F_1'(z)-F_2'(z)=f(z)-f(z)=0$$

at every point of D. According to the Cauchy-Reimann conditions (19.1), we have in D:

$$\frac{\partial U}{\partial x}=\frac{\partial V}{\partial y}=0 \text{ and } \frac{\partial V}{\partial x}=-\frac{\partial U}{\partial y}=0$$

from which it follows that the functions U and V are constants. Consequently, G(z) is also a constant function.

19.3.1 Example: Indefinite integral of the inverse function

The function $f(z)=1/z$ is analytic except at the origin (fig. 19.6).

In a simply connected region that does not contain the origin, the line integral:

$$I(z)=\int_a^z \frac{dz}{z}$$

thus is a single-valued analytic function. Because

$$\frac{1}{z}=\frac{1}{x+jy}=\frac{x-jy}{x^2+y^2}$$

we have

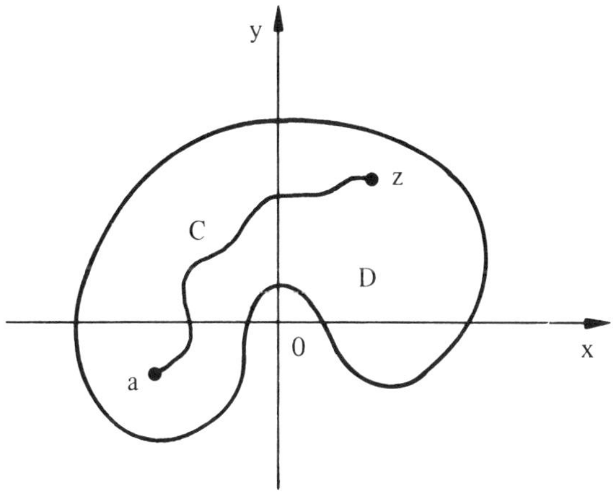

Fig. 19.6

$$I(z) = \int_C \frac{dz}{z} = \int_C \frac{(x - jy)(dx + j\,dy)}{x^2 + y^2}$$

$$= \int_C \frac{x\,dx + y\,dy}{x^2 + y^2} + j \int_C \frac{x\,dy - y\,dx}{x^2 + y^2}$$

To calculate these real line integrals, we set $z = re^{j\varphi}$, and we find:

$$I(z) = \int_C \frac{dr}{r} + j \int_C d\varphi = \ln \left| \frac{z}{a} \right| + j \arg \frac{z}{a} = \ln \frac{z}{a} = \ln z - \ln a$$

Chapter 20

Cauchy's Integral Formula and Applications

20.1 CAUCHY'S INTEGRAL FORMULA

The most important consequence of Cauchy's integral theorem is Cauchy's integral formula, which we will derive in the following.

To start with, let f(z) be an analytic function in a simply connected region D and on its boundary C (fig. 20.1). Then, we consider the function g(z):

$$g(z) = \frac{f(z)}{z-a}$$

where z = a is an ***interior*** point of D. Evidently, the function g(z) is analytic in D and on its contour C, with the exception of the point z = a. Eliminating from the domain D a disk of radius ϵ centered at the point z = a and making a "crosscut" from the curve C to the boundary Γ_ϵ of this disk, we obtain a simply connected region D* (fig. 20.2) in which g(z) is analytic. According to Cauchy's theorem the line integral of g(z) along the complete boundary C* of D* is zero:

$$\oint_{C^*} g(z)\,dz = \oint_{C^*} \frac{f(z)}{z-a}\,dz = 0 \qquad (20.1)$$

As the width of the "crosscut" is decreased toward zero, the integrals along the edges are taken in opposite directions and hence cancel. Consequently, equation (20.1) can be written as follows:

$$\oint_{C} \frac{f(z)}{z-a}\,dz + \oint_{\Gamma_\epsilon} \frac{f(z)}{z-a}\,dz$$

and changing the direction of integration along the circle Γ_ϵ, it follows that

$$\oint_{C} \frac{f(z)}{z-a}\,dz = \oint_{-\Gamma_\epsilon} \frac{f(z)}{z-a}\,dz \qquad (20.2)$$

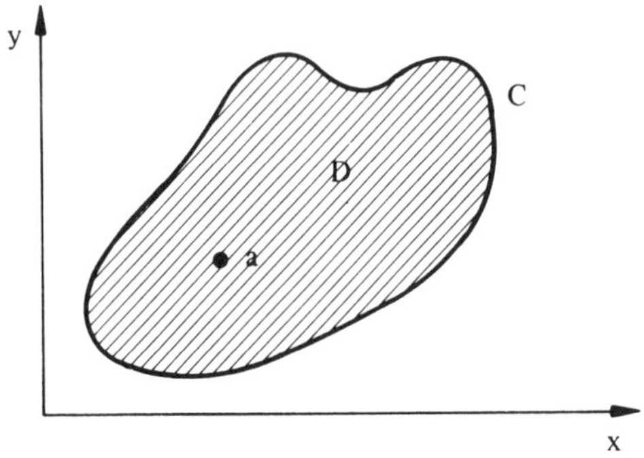

Fig. 20.1

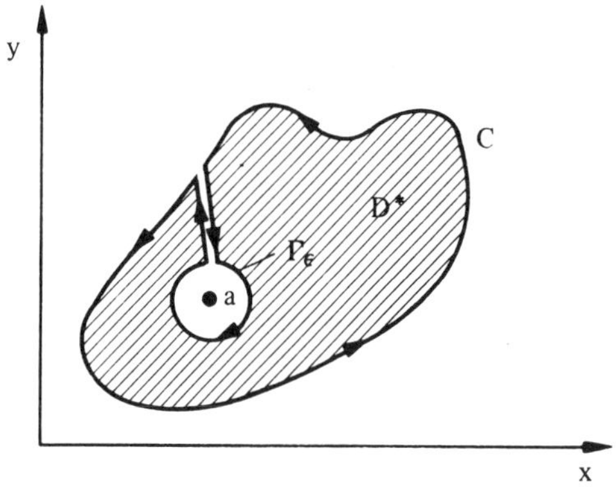

Fig. 20.2

The parametric representation of the circle Γ_ϵ is

$$z = a + \epsilon\, e^{j\varphi}, \quad 0 \leq \varphi \leq 2\pi$$

and

$$dz = j\epsilon\, e^{j\varphi}\, d\varphi$$

The integral on the right side of relation (20.2) becomes

$$\oint_{-\Gamma_\epsilon} \frac{f(z)}{z-a}\, dz = \int_0^{2\pi} \frac{f(a+\epsilon\, e^{j\varphi})\, j\epsilon\, e^{j\varphi}}{\epsilon\, e^{j\varphi}}\, d\varphi = j \int_0^{2\pi} f(a+\epsilon e^{j\varphi})\, d\varphi$$

In the limit, as $\epsilon \to 0$, we find, taking into account that $f(z)$ is a continuous function:

$$\oint_C \frac{f(z)}{z-a}\,dz = 2\pi j\, f(a)$$

or

$$f(a) = \frac{1}{2\pi j}\oint_C \frac{f(z)}{z-a}\,dz \tag{20.3}$$

which is ***Cauchy's integral formula***. Differentiating the left and right sides of equation (20.3) with respect to the variable a, we obtain for the first derivative:

$$f'(a) = \frac{1}{2\pi j}\oint_C \frac{f(z)}{(z-a)^2}\,dz$$

and, in general, we find for the *n*th derivative:

$$f^{(n)}(a) = \frac{n!}{2\pi j}\oint_C \frac{f(z)}{(z-a)^{n+1}}\,dz, \quad n = 1, 2, \ldots \tag{20.4}$$

Cauchy's integral formula is quite remarkable because it shows that if an analytic function is known on a closed contour, then it is also known within this contour. Further, the formulas for the derivatives show that if a function of a complex variable is analytic, i.e., it has a first derivative, then all the higher derivatives exist as well. This, of course, is not true for functions of real variables.

Finally, we note that if $z = a$ is a point ***exterior*** to D, the function:

$$g(z) = \frac{f(z)}{z-a}$$

is analytic in the interior of D and on its contour C. In this case, we have from Cauchy's theorem:

$$\oint_C \frac{f(z)}{z-a}\,dz = 0$$

20.1.1 Example: Evaluation of a line integral using Cauchy's formula

Evaluate the line integral:

$$\oint_C \frac{dz}{z^2 + a^2}$$

where C is a circle of radius $r > |a|$ centered at the origin.

The function:

$$h(z) = \frac{1}{z^2 + a^2}$$

is analytic everywhere except at the two points $z_1 = ja$ and $z_2 = -ja$, which are located inside the circle $|z| = r > |a|$ (fig. 20.3).

Expanding $h(z)$ into partial fractions:

$$h(z) = \frac{1}{z^2 + a^2} = \frac{1}{(z+ja)(z-ja)} = \frac{1}{2ja}\left[\frac{1}{z-ja} - \frac{1}{z+ja}\right]$$

we find, applying the Cauchy integral formula (20.3), for the analytic function $f(z) \equiv 1$:

$$\oint_C \frac{dz}{z^2+a^2} = \frac{1}{2ja}\left[\oint_C \frac{dz}{z-ja} - \oint_C \frac{dz}{z+ja}\right] = \frac{1}{2ja}(2\pi j - 2\pi j) = 0$$

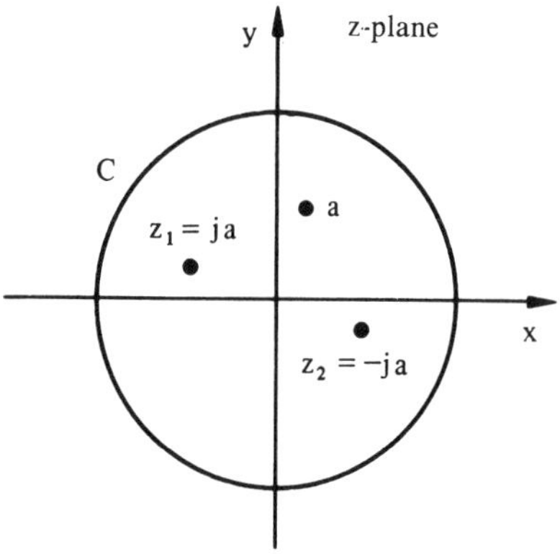

Fig. 20.3

20.1.2 Example: Mean value of a periodic function

Determine the mean value $\overline{g}$ of the periodic function of period 2π:

$$g(\varphi) = \frac{1}{\frac{\sqrt{5}}{2} + \cos\varphi}$$

which is defined by

$$\overline{g} = \frac{1}{2\pi}\int_0^{2\pi}\frac{d\varphi}{\frac{\sqrt{5}}{2}+\cos\varphi} \tag{20.5}$$

The transformation $z = e^{j\varphi}$ maps the integration interval $[0, 2\pi]$ onto the unit circle in the complex z-plane. We have

$$dz = je^{j\varphi}d\varphi = jz\,d\varphi$$

and

$$\cos\varphi = \frac{1}{2}(e^{j\varphi}+e^{-j\varphi}) = \frac{1}{2}\left(z+\frac{1}{z}\right)$$

After substituting these relations into the integral (20.5), the average $\overline{g}$ becomes

$$\overline{g} = \frac{1}{2\pi}\oint_C \frac{dz}{j\frac{z}{2}\left(\sqrt{5}+z+\frac{1}{z}\right)} = \frac{1}{\pi j}\oint_C \frac{dz}{z^2+\sqrt{5}z+1}$$

where the path of integration C is the unit circle centered at the origin.

The roots of the polynomial of the denominator are

$$z_1 = \frac{-\sqrt{5}+1}{2} \text{ and } z_2 = -\frac{\sqrt{5}+1}{2}$$

where z_1 is located inside and z_2 outside the unit circle. Thus, we obtain

$$\overline{g} = \frac{1}{j\pi}\oint_C \frac{dz}{(z-z_1)(z-z_2)} = \frac{1}{j\pi}\oint_C \frac{f(z)}{z-z_1}dz$$

where the function:

$$f(z) = \frac{1}{z-z_2}$$

is analytic within and on the unit circle, and using Cauchy's integral formula (20.3), we find

$$\overline{g} = \frac{2\pi j}{j\pi}f(z_1) = 2\frac{1}{z_1-z_2} = 2$$

20.1.3 Example: Integral of a power function

Let us consider the closed line integral:

$$\oint_C \frac{dz}{(z-a)^n}, \quad \text{n positive integer}$$

where C designates a simple closed curve enclosing the point $z = a$.

Defining $f(z) \equiv 1$, we obtain from Cauchy's formula and the formula for its derivatives:

$$\oint_C \frac{dz}{(z-a)^n} = \begin{cases} 2\pi j & \text{if } n = 1 \\ 0 & \text{if } n = 2, 3, 4 \ldots \end{cases} \tag{20.6}$$

If $n = 0, -1, -2, \ldots$ the integrand is analytic everywhere inside the simple closed curve, including $z = a$. Hence, by Cauchy's theorem, the integral is zero.

20.2 TAYLOR SERIES

The familiar Taylor series is an effective tool in real calculus and its applications. We shall now see that every analytic function can be represented by power series which are quite similar to the Taylor series of real functions.

Let $f(z)$ be an analytic function inside and on the circle C of radius $r > 0$ centered at $z = b$. At an arbitrary point $z = a$ at the interior of C, we have according to the Cauchy integral formulas (20.3) and (20.4):

$$f^{(n)}(a) = \frac{n!}{2\pi j} \oint_C \frac{f(z)}{(z-a)^{n+1}} dz, \quad n = 0, 1, 2, \ldots$$

We now consider the following expansion:

$$\frac{1}{z-a} = \frac{1}{z-b-(a-b)} = \frac{1}{z-b} \frac{1}{1 - \dfrac{a-b}{z-b}}$$

$$= \frac{1}{z-b} \left[1 + \frac{a-b}{z-b} + \frac{(a-b)^2}{(z-b)^2} + \ldots \right] \tag{20.7}$$

and since the point z is on the contour C, we have $|z-b| > |a-b|$, i.e., the geometric progression in brackets on the right side of relation (20.7) converges (fig. 20.4).

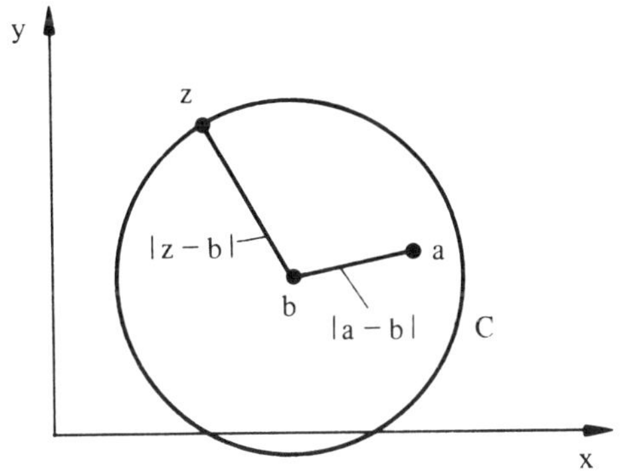

Fig. 20.4

Substituting the term $1/(z-a)$ in Cauchy's formula (20.3):

$$f(a) = \frac{1}{2\pi j} \oint_C \frac{f(z)}{z-a}\, dz$$

by the geometric progression (20.7), we obtain

$$f(a) = \frac{1}{2\pi j} \oint_C \frac{f(z)}{z-b}\, dz + \frac{a-b}{2\pi j} \oint_C \frac{f(z)}{(z-b)^2}\, dz$$

$$+ \frac{(a-b)^2}{2\pi j} \oint_C \frac{f(z)}{(z-b)^3}\, dz + \ldots.$$

from which we obtain, again by using the Cauchy integral formula (20.4), the ***Taylor series*** of the analytic function $f(z)$ with center at the point $z = b$.

$$f(a) = f(b) + f'(b)(a-b) + \frac{f''(b)}{2!}(a-b)^2 + \ldots \tag{20.8}$$

This Taylor-series expansion is valid for all the points interior to the circle C.

20.2.1 Example: Taylor-series expansion of the sine function

Expand the function:

$$f(z) = \sin z$$

in a Taylor series with center at the point $z = \pi/4$.

Again designating the variable by z, formula (20.8) takes the following form:

$$f(z) = f(\pi/4) + f'(\pi/4)(z - \pi 4) + \frac{f''(\pi/4)}{2!}(z - \pi/4)^2 + \ldots$$

which gives

$$f(z) = \frac{\sqrt{2}}{2}\left\{ 1 + (z - \pi/4) - \frac{(z - \pi/4)^2}{2!} - \frac{(z - \pi/4)^3}{3!} + \ldots \right\}$$

Chapter 21

Integration by the Method of Residues

21.1 SINGULARITIES AND THEIR RESIDUES

Let f(z) be a single-valued and analytic function within and on a simple closed curve C, except at the isolated points a, b, c, . . , interior to C (fig. 21.1)

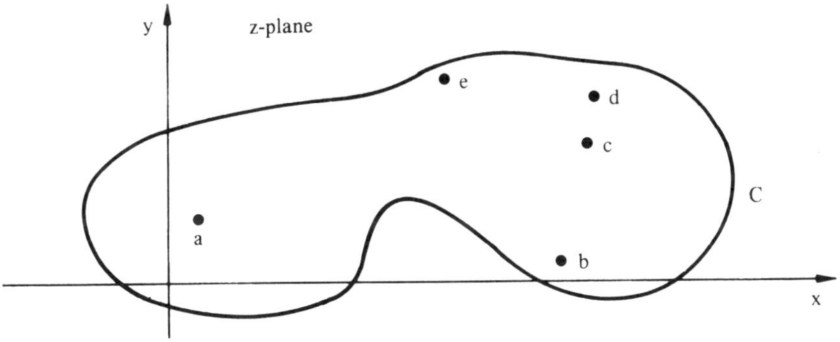

Fig. 21.1

Analytic functions may have different types of singularities. First, we remember that a singular point of an analytic function f(z) is a point where f(z) ceases to be analytic. Then, we say that f(z) has a singularity at that point. ***Poles*** are by definition isolated singularities. If a single-valued and analytic function has a singularity other than a pole, this singularity is called an ***essential singularity*** of f(z).

In the following, we shall only consider poles, although the results obtained are also valid for essential singularities.

An isolated singularity z = a of the function f(z) is called an ***nth-order pole*** of f(z), if f(z) can be written in the neighborhood of z = a in the form:

$$f(z)=g(z)\ \frac{1}{(z-a)^n} \tag{21.1}$$

where n is a positive integer and g(z) is an analytic function in the neighborhood of z = a, such that $g(a)\neq 0$.

In this case, we can expand g(z) in a Taylor series with center at the point z = a:

$$g(z)=g(a)+g'(a)(z-a)+g''(a)\ \frac{(z-a)^2}{2!}+\ldots$$

and relation (21.1) becomes

$$f(z) = \frac{g(a)}{(z-a)^n} + \frac{g'(a)}{(z-a)^{n-1}} + \ldots + \frac{g^{(n-1)}(a)}{(n-1)!(z-a)} + \frac{g^{(n)}(a)}{n!}$$

$$+ \frac{g^{(n+1)}(a)}{(n+1)!}(z-a)+\ldots = \sum_{k=-n}^{\infty} \alpha_k (z-a)^k \tag{21.2}$$

with

$$\alpha_k = \frac{g^{(n+k)}(a)}{(n+k)!}, \quad k=-n,\ -n+1,\ldots$$

The right side of (21.2) is called a ***Laurent series***, and the part containing the negative powers of a Laurent series is called its ***principal part***. It is clear that a Laurent series is reduced to a Taylor series if its principal part vanishes.

The coefficient α_{-1} of the first negative power in the representation of f(z) by a Laurent series (21.2) is called the ***residue*** of f(z) at z = a. It is given by

$$\alpha_{-1} = \frac{g^{(n-1)}(a)}{(n-1)!} = \frac{1}{(n-1)!}\ \frac{d^{n-1}}{dz^{n-1}}\{(z-a)^n f(z)\}_{z=a} \tag{21.3}$$

In particular, we have for a simple pole (a first-order pole):

$$\alpha_{-1}=\{(z-a)f(z)\}_{z=a}$$

and for a second-order pole:

$$\alpha_{-1} = \frac{d}{dz}\{(z-a)^2 f(z)\}_{z=a}$$

Let us consider the example:

$$f(z) = \frac{z}{(z-1)(z+1)^2}$$

Then, z = 1 and z = −1 are respectively the poles of order 1 and order 2. For the residue at z = 1, we have

$$\{(z-1)\,f(z)\}_{z=1} = \left\{ \frac{z}{(z+1)^2} \right\}_{z=1} = \frac{1}{4}$$

and for the residue at z = –1:

$$\frac{d}{dz}\{(z+1)^2\, f(z)\}_{z=-1} = \frac{d}{dz}\left\{ \frac{z}{z-1} \right\}_{z=-1} = -\frac{1}{4}$$

21.2 RESIDUE THEOREM

Let f(z) be a ***single-valued*** and ***analytic*** function inside and on a simple closed curve C, except at a finite number of singularities a,b,c, . . . , interior to C in which the residues of f(z) are respectively α_{-1}, β_{-1}, γ_{-1}, . . . Then, the ***residue theorem*** states that:

$$\oint_C f(z)\,dz = 2\pi j(\alpha_{-1} + \beta_{-1} + \gamma_{-1} + \ldots) \tag{21.4}$$

To demonstrate this theorem, we enclose each singularity in a circle with a sufficiently small radius (fig. 21.2) so as to be completely inside C. Thus, according to Cauchy's theorem, we have

$$\oint_C f(z)\,dz + \oint_{C_a} f(z)\,dz + \oint_{C_b} f(z)\,dz + \ldots = 0$$

considering the fact that the integrals along the edges of the "crosscuts" are taken in opposite directions and hence cancel as the width is reduced to zero.

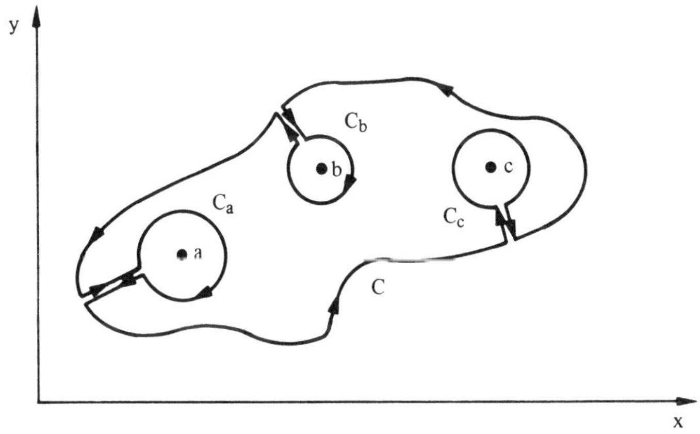

Fig. 21.2

In the neighborhood of the singularity $z = a$, we may represent the function $f(z)$ by its Laurent series:

$$f(z) = \sum_{k=-\infty}^{\infty} \alpha_k (z-a)^k \tag{21.5}$$

and according to equation (20.6)

$$\oint_{C_a} (z-a)^k \, dz = \begin{cases} 0 & \text{for } k \neq -1 \\ -2\pi j & \text{for } k = -1 \end{cases}$$

we obtain for the line integral of $f(z)$ along the circle C_a by integrating the right side of relation (21.5) term by term:

$$\oint_{C_a} f(z)\, dz = \sum_{k=-\infty}^{\infty} \alpha_k \oint_{C_a} (z-a)^k \, dz = -2\pi j \, \alpha_{-1}$$

This integral, therefore, equals the single coefficient α_{-1}, which explains the importance attached to the residue of a function at a singularity. Analogously, we find the value $-2\pi j \, \beta_{-1}$ for the integral of $f(z)$ along the circle C_b and so on. Consequently, the ***closed line integral***:

$$\oint_{C} f(z)\, dz$$

is ***equal*** to $2\pi j$ ***times the sum of the residues at all the singularities located inside the simple closed curve*** C.

21.3 APPLICATION

The residue theorem yields a very elegant and simple method for evaluating certain classes of complicated real integrals. In many cases the integration can be made a routine matter as illustrated by the following example where we are to determine the mean value of a periodic function:

$$I = \frac{1}{2\pi} \int_0^{2\pi} \frac{d\varphi}{25 - 16\cos^2\varphi}$$

Using the transformation $z = \exp(j\varphi)$ we map the segment $[0, 2\pi]$ onto the unit circle in the complex plane of the variable z. Because

$$\cos\varphi = \frac{e^{j\varphi} + e^{-j\varphi}}{2} = \frac{1}{2}\left(z + \frac{1}{z}\right)$$

and

$$dz = je^{j\varphi}\,d\varphi = jz\,d\varphi$$

the given real integral I is transformed into a closed line integral along the unit circle:

$$I = -\frac{j}{2\pi}\oint \frac{dz}{z[25-4(z+1/z)^2]}$$

or, simplifying,

$$I = \frac{j}{2\pi}\oint \frac{z}{4z^2-17z^2+4}\,dz$$

The function to be integrated has ***simple poles*** at $z_{1,2} = \pm 2$, $z_{3,4} = \pm 1/2$, of which only $z_{3,4} = \pm 1/2$ are interior to the unit circle. The residue at the pole $z_3 = 1/2$ is given by formula (21.3) for $n = 1$:

$$\left\{(z-1/2)\frac{z}{4(z^2-4)(z-1/2)(z+1/2)}\right\}_{z=1/2} = -\frac{1}{30}$$

and the residue at $z_4 = -1/2$ is given by

$$\left\{(z+1/2)\frac{z}{4(z^2-4)(z-1/2)(z+1/2)}\right\}_{z=-1/2} = -\frac{1}{30}$$

From the residue theorem, we obtain

$$I = \frac{1}{2\pi}\int_0^{2\pi}\frac{d\pi}{25-16\cos^2\varphi} = \frac{2\pi j^2}{2\pi}\left(-\frac{1}{30}-\frac{1}{30}\right) = \frac{1}{15}$$

Bibliography

1. Y. Baranger, *Introduction à l'Analyse Numérique*, Hermann, Paris, 1977.
2. J. Bass, *Cours de mathématiques, vols. 1 & 2, Masson, Paris, 1968.*
3. W. Böhm and G. Gose, *Einführung in die Methoden der numerischen Mathematik*, Vieweg, Braunschweig, 1977.
4. G. Charet, *Cours d'Analyse Numérique*, Société d'Enseignement supérieur, Paris, 1975.
5. L. Collatz and J. Albrecht, *Aufgaben aus der angewandten Mathematik*, vols. 1 & 2, Vieweg, Braunschweig, 1973.
6. S.D. Conte and C. de Boor, *Elementary Numerical Analysis*, McGraw-Hill, New York, 1972.
7. M. Decuyper, *Compléments de mathématiques*, Dunod, Paris, 1968.
8. J. M. Escané, *Fonctions d'une variable complexe*, Masson, Paris, 1972.
9. C.E. Fröberg, *Introduction to Numerical Analysis*, Addison-Wesley, London, 1973.
10. R.W. Hamming, *Numerical Methods for Scientists and Engineers*, International Student Edition, McGraw-Hill, New York, 1973.
11. P. Henrici, *Elements of Numerical Analysis*, John Wiley and Sons, New York, 1964.
12. P. Henrici and R. Jeltsch, *Komplexe Analysis für Ingenieure*, vols. 1 & 2, Birkhäuser, 1977, 1980.
13. F.B. Hildebrand, *Introduction to Numerical Analysis*, McGraw-Hill, New York, 1956.

14. M. Lavrentiev and B. Chabat, *Méthodes de la théorie des fonctions d'une variable complexe*, MIR, Moscow, 1977.

15. T.R. McCalla, *Introduction to Numerical Methods and Fortran Programming*, John Wiley and Sons, New York, 1967.

16. J. Ortusi, *Mathématiques appliquées à l'électronique*, vols. 1 & 2, Dunod, Paris, 1969.

17. N. Piskounov, *Calcul différentiel et intégral*, vols. 1 & 2, MIR, Moscow, 1980.

18. J. Quinet, *Cours élémentaires de mathématiques supérieures,* vols. 3 & 4, Dunod, Paris, 1977.

19. F. Scheid, *Numerical Analysis*, Schaum's Outline Series, 1968.

20. V. Smirnov, *Cours de mathématiques supérieures*, vols. 2, 3, 4, MIR, Moscow, 1972, 1975.

21. M.R. Spiegel, *Analyse vectorielle*, McGraw-Hill, Paris, 1977.

22. M.R. Spiegel, *Théorie et applications de l'analyse*, McGraw-Hill, Paris, 1973.

23. M.R. Spiegel, *Analyse de Fourier*, McGraw-Hill, Paris, 1980.

24. M.R. Spiegel, *Transformées de Laplace*, McGraw-Hill, Paris, 1980.

25. E. Stiefel, *Introduction à la mathématique numérique*, Dunod, Paris, 1967.

26. J. Stoer, *Einführung in die numerische Mathematik*, vols. 1 & 2, Springer Verlag, Berlin, 1976.

27. I. Zeldovitch and A. Mychkis, *Eléments de mathématiques appliquées*, MIR, Moscow, 1974.

Index